Springer Proceedings in Energy

The series Springer Proceedings in Energy covers a broad range of multidisciplinary subjects in those research fields closely related to present and future forms of energy as a resource for human societies. Typically based on material presented at conferences, workshops and similar scientific meetings, volumes published in this series will constitute comprehensive state-of-the-art references on energy-related science and technology studies. The subjects of these conferences will fall typically within these broad categories:

- Energy Efficiency
- Fossil Fuels
- Nuclear Energy
- Policy, Economics, Management & Transport
- Renewable and Green Energy
- Systems, Storage and Harvesting
- Materials for Energy

eBook Volumes in the Springer Proceedings in Energy will be available online in the world's most extensive eBook collection, as part of the Springer Energy eBook Collection. To submit a proposal or for further inquiries, please contact the Springer Editor in your region:

Kamiya Khatter (India)
Email: kamiya.khatter@springer.com
Loyola D'Silva (All other countries)
Email: loyola.dsilva@springer.com

More information about this series at http://www.springer.com/series/13370

R. Seyezhai · S. Karuppuchamy · L. Ashok Kumar
Editors

Recent Trends in Renewable Energy Sources and Power Conversion

Select Proceedings of ICRES 2020

Springer

Editors
R. Seyezhai
Sri Sivasubramaniya Nadar College
of Engineering
Kalavakkam, Chennai, Tamilnadu, India

S. Karuppuchamy
Alagappa University
Karaikudi, India

L. Ashok Kumar
PSG College of Technology
Coimbatore, India

ISSN 2352-2534 ISSN 2352-2542 (electronic)
Springer Proceedings in Energy
ISBN 978-981-16-0668-7 ISBN 978-981-16-0669-4 (eBook)
https://doi.org/10.1007/978-981-16-0669-4

This Springer imprint is published by the registered company Springer Nature Singapore Pte Ltd.
The registered company address is: 152 Beach Road, #21-01/04 Gateway East, Singapore 189721, Singapore

Preface

Renewable energy sources have finally quenched the thirst for the sempiternal search for an alternative clean and green energy source that could pave way for a promising, sustainable future. A myriad of renewable sources like solar, wind, ocean, biomass, and fuel cells have been delineated carefully in this book. The intermittent nature of these sources makes power electronic converters an indispensable part of the renewable energy setup. Several power converter topologies like AC–DC, DC–DC, DC–AC, and AC–AC are employed to achieve notable efficiency. Advanced power semiconductor devices, their gate drive and protection circuits, heat sink design, and magnetic components for power converter are some of the additional topics that are discussed in this book.

This book is a compilation of full-length papers presented during the proceedings of the First Virtual International Conference on Renewable Energy Systems (ICRES 2020), held during August 26–28, 2020, organized by the Department of Electrical and Electronics Engineering, Sri Sivasubramaniya Nadar College of Engineering, Kalavakkam. The substantive areas of focus of the conference were the various viable methodologies and innovations that would help in the shift toward renewable sources in the future, which could help create a potentially huge impact among academicians, researchers, policy-makers, scientists, practitioners, and students with respect to this domain. The key topics covered in this book include solar photovoltaic system, wind energy, biomass, modern power converter topologies, and application of IoT for energy sources.

The conference garnered around over 60 registrations, which also includes national as well as international participation. The paper presentations were from several gilt-edged institutions like IITs, NITs, IIITDM, and Anna University. The three-day event had spiffing lectures from five keynote speakers that articulated various ideologies related to the domain. The papers included in this book have been reviewed by the domain experts to ensure the quality and correctness of the manuscript. ICRES 2020 technical committee and editors thank the authors for their support during the review process. This book will serve as a one-stop reference comprehensively explaining the various renewable energy sources, technological development, and advanced power conversion topologies.

Chennai, India R. Seyezhai
Karaikudi, India S. Karuppuchamy
Coimbatore, India L. Ashok Kumar
August 2020 Volume Editors, ICRES 2020

Contents

About the Editors

Dr. R. Seyezhai is Associate Professor in the Department of Electrical and Electronics Engineering, SSN College of Engineering, Chennai. She has 21 years of teaching experience including 14 years of research experience. She obtained her B.E. degree in Electronics and Communication Engineering and M.E. degree in Power Electronics and Drives from Manonmaniam Sundaranar University, Tirunelveli, and Bharathidasan University, Tiruchirapalli, respectively, and Ph.D. in the area of Multilevel Inverters for Fuel Cell Systems from Anna University, Chennai. During her Ph.D., she developed a prototype of SiC based hybrid cascaded multilevel inverter employing a novel modulation technique for fuel cell power conditioning system (PCS). She has published 300 research publications in international journals/ international and national conferences. She has published one book and one book chapter. She received best teacher award for the academic years 2004–2005, 2013–2014, 2014–2015, 2015–2016, 2016–2017, 2017–2018 and the CTS best faculty award for the year 2012–2013. She is recipient of "Bharat Jyothi award 2012" from the International Friendship society for outstanding contribution to student's community, "Best researcher Award 2015" by ASDF, "Outstanding Researcher Award 2016 in Power Electronics" by Venus International foundation and "Research Excellence Award 2017" by Institute for exploring advances in engineering. Ten students completed their Ph.D. and currently she is guiding five Ph.D. students including four full-time candidates in the area of renewable energy system and its interface circuits. Currently, she is working on AICTE MODROBS project for the modernization of Renewable Energy Conversion Laboratory for Rs. 11 lakhs during 2020–2021. She has completed AICTE research projects titled 'Design of silicon carbide based hard switched DC–AC power converter' during 2007–2010 and 'Design and development of Z-source cascaded multilevel inverter for photovoltaic applications' during 2013–2016. Also, she completed the project 'Development of solar powered electric vehicle using BLDC drive' funded by SSN Trust and an internal funded project titled 'Quasi Z-source Multilevel Inverter for PV applications'. She has filed two patents. She has developed 'Renewable Energy Conversion Laboratory' at SSNCE with the help of external and internal funding. She is in the editorial board and reviewer board of several international journals. She is an active member in IEEE and life member in ISTE. Currently, she is the chairman, IEEE-PELS, Madras Section.

Dr. S. Karuppuchamy is Professor and Head, Department of Energy Science, Alagappa University, Karaikudi. He has 23 years of teaching/research experience at various international universities and industries. He has received his Ph.D. from Gifu University, Gifu, Japan and D.Sc. from Periyar University, India. His current research interests include the development of energy conversion and storage devices. He has received funding for research and development from various funding agencies to the tune of 5.30 crores. He has published 147 articles in reputed international journals. To his credit, he has 11 international patents in the field of chemical sciences. He has presented 170 research papers in international and national conferences. He received more than 20 best paper awards from various international and national conferences. He has organized six conferences and delivered more than 85 invited lectures in various international and national institutes/conferences. He has authored nine books and four book chapters. Dr. S. Karuppuchamy's research contribution has more than 2414 citations to his credit. His h-index is 32 and i10-index is 58. He is serving as editorial board member for various international journals. Moreover, he is a life member of Chemical Research Society of India; Society for Advancement of Electrochemical Science and Technology, India; International Society of Electrochemistry, Switzerland; Asian Federation of Biotechnology, South Korea; and American Nano Society, USA. He is the recipient of various prestigious international awards viz, MONBUSHO, AsiaBiomass Energy Researcher Award, New Energy Foundation and Invited Researcher Award, National Energy Development Organization (NEDO) from Government of Japan, Innovation Award from Henkel KGaA, Germany, Scientific Adviser award from TSM Co. Ltd., South Korea and SPD Laboratory Co. Ltd., Japan. He received Young Scientist award from DST, Government of India. In 2017, he was awarded Alagappa Excellence Award for his research contributions. In recognition of the outstanding research work done by Dr. S. Karuppuchamy in the field of Nanomaterials and Photoelectrochemistry the Tamil Nadu State Council for Science and Technology awarded him the Tamil Nadu Scientist Award (TANSA) in Chemical Sciences for the year 2017. Recently, he has become the Fellow of Academy of Sciences Chennai (FASC).

Dr. L. Ashok Kumar was a Postdoctoral Research Fellow from San Diego State University, California. He is a recipient of the BHAVAN fellowship from the Indo-US Science and Technology Forum and SYST Fellowship from DST, Government of India. He has 3 years of industrial experience and 20 years of academic and research experience. He has published 173 technical papers in International and National journals and presented 167 papers in National and International Conferences. He has completed 26 Government of India funded projects worth about 15 Crores and currently seven projects are in progress worth about 7 Crores. He has developed 27 products and out of that 14 products have been technology transferred to industries and for Government funding agencies. His Ph.D. work on wearable electronics earned him a National Award from ISTE, and he has received 24 awards on the National level. He has guided 92 graduate and postgraduate projects. He has produced five Ph.D. Scholars and 12 candidates are doing Ph.D. under his supervision. He has visited many countries for institute industry collaboration and as a keynote speaker. He has been an invited speaker in 235 programs. Also he has organized 102 events, including conferences, workshops, and seminars. He completed his graduate program in Electrical and Electronics Engineering from University of Madras and his post-graduate from PSG College of Technology, India, and Masters in Business Administration from IGNOU, New Delhi. After completion of his graduate degree, he joined as project engineer for Serval Paper Boards Ltd., Coimbatore (now ITC Unit, Kovai). Presently he is working as a Professor and Associate HoD in the Department of EEE, PSG College of Technology. He is also a Certified Charted Engineer and BSI Certified ISO 500001 2008 Lead Auditor. He has authored 14 books in his areas of interest published by Springer, USA, CRC Press, USA, Cambridge University Press, London, Elsevier Press, London and Wiley India and he also have 11 patents to his credit and also contributed 18 chapters in various books. He is also the Chairman of Indian Association of Energy Management Professionals and Joint Secretary of Institution of Engineers, Coimbatore. He is holding prestigious positions in various national and international forums and he is a Fellow Member in IET (UK), Fellow Member in IETE, Fellow Member in IE and Senior Member in IEEE.

Chapter 1
Simulation and Analysis of Controllers for Bridgeless Interleaved AC–DC SEPIC PFC Converter

Sasikala Rajagopal and R. Seyezhai

Abstract In this paper, a new bridgeless interleaved SEPIC converter is investigated for LED applications. LEDs require a stable and good quality power supply for an extended lifetime with reduced flicker. The converter output is affected by supply and load side disturbances. This will lead to increased flicker and it will reduce the life span of LEDs. To alleviate this problem, a controller is implemented for the proposed converter to maintain a constant output voltage. Both proportional integral (PI) and fuzzy logic controller (FLC) are implemented, and performance of the proposed converter was analyzed based on different parameters such as overshoot voltage, rise time, settling time, steady-state error, supply PF, supply current THD. The converter circuit along with the two different controllers is studied and simulated with the help of software MATLAB/Simulink. The FLC controller gives constant output voltage for change in load. The results are verified.

Keywords PFC converter · Voltage mode control · PI controller · Fuzzy logic controller

1.1 Introduction

Light emitting diode (LED) lighting is mostly used in recent years because of their eco-friendly, durability, and maintenance-free requirement [1, 2]. They are widely used in domestic and industrial applications [3–5]. AC to DC PFC boost converter is commonly used for LED drivers [6–8]. But boost PFC rectifier has poor power factor and supply current is also rich in harmonics. Moreover, with the interleaving technique applied to boost converter, the supply current harmonics does not satisfy

S. Rajagopal (✉)
Sathyabama Institute of Science and Technology, Chennai, India

R. Seyezhai
Sri Sivasubramaniya Nadar College of Engineering, Kalavakkam, Chennai, Tamilnadu, India
e-mail: seyezhair@ssn.edu.in

R. Seyezhai et al. (eds.), *Recent Trends in Renewable Energy Sources and Power Conversion*, Springer Proceedings in Energy,
https://doi.org/10.1007/978-981-16-0669-4_1

the IEC standards. To improve the power factor, this paper focuses on the interleaved SEPIC converter which results in better power factor correction [9, 10].

In conventional interleaved SEPIC model, the load said voltage is unregulated. This unregulated load voltage may cause flickering problem in LED. So we have to regulate the voltage at the load side to avoid LED flickering. Various control methods are applied her for the regulation of output voltage. In [11, 12], the authors have proposed a new converter in which the voltage at the load side is controlled by using a PI controller [13, 14]. But power converters are nonlinear in nature; hence, the FLC method is proposed in this paper. The fuzzy logic controller has finer voltage regulation comparably good with respect to other control techniques in [15]. Both PI controller and FLC are designed in the bridgeless Interleaved SEPIC PFC converter. The performance of the bridgeless interleaved SEPIC PFC converter was analyzed based on the different parameters to find the best one for real-time applications. From the output, it is found that the FLC-based controller gives a better regulation and power factor than the PI controller.

The paper is ordered as follows: Sect. 1.2 gives description about interleaved SEPIC PFC converter. Section 1.3 discusses the controllers employed for the power converter circuit. Section 1.4 gives the simulation results and comparison. Sections 1.5 and 1.6 analyze about results and conclude paper.

1.2 Description

The circuit configuration of interleaved bridgeless SEPIC converter is shown in Fig. 1.1. This configuration is used to lower the supply current ripple and improve the input power factor. The circuit has two different modes of operation. During mode-1, switches Q_1 and Q_3 Conduct; at that time, the L_1 and L_3 get charged, and L_2 and L_4 discharge the energy through capacitor C_2 and C_4 via body the diode of

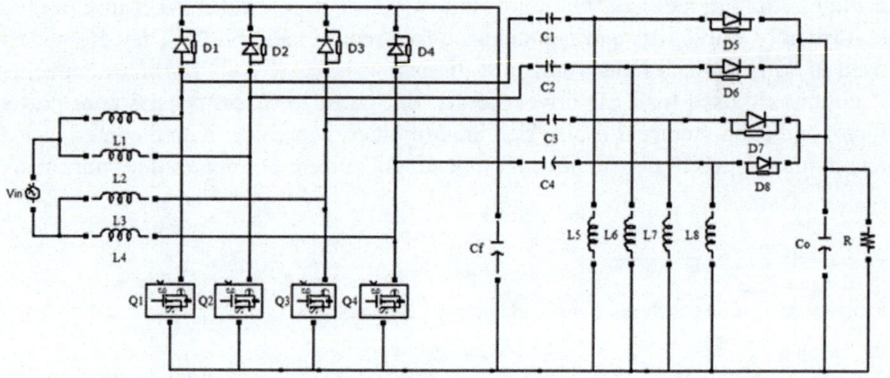

Fig. 1.1 Circuit diagram of bridgeless IBC SEPIC converter

Table 1.1 Simulation parameters

Supply voltage	$V_{in} = 24$ V
Load voltage	$V_{out} = 48$ V
Input frequency	50 Hz
Switching frequency	50 kHz
Duty cycle	0.5
Inductor	$L_1, L_2, L_3, L_4 = 0.35$ mH $L_5, L_6, L_7, L_8 = 79$ µH
Capacitor	$C_1, C_2, C_3, C_4 = 1.05$ µF $C_0 = 4700$ µF

Q_4/Q_2 switches, and this energy is distributed to the load. L_5 and L_7 are connected to the switches Q_1 and Q_3 on load side. The take energy from the capacitors C_1 and C_3 through the output capacitor is supplied to the load. During mode-2, switches Q_2 and Q_4 conduct; at that time, the L_2 and L_4 get charged, and L_1 and L_3 discharge the energy through capacitor C_1 and C_3 via body diode of Q_1/Q_3 switches, and this energy is delivered to the load [16]. L_6 and L_8 are connected to the switches Q_2 and Q_4 takes energy from the output capacitors C_2 and C_4 and the energy is supplied to the load.

Design equations for the proposed converter:

$$\text{Output voltage} \quad V_O = \frac{DV_{in}}{(1 - D)} \tag{1.1}$$

$$\text{Boost inductor} \quad L_1, L_2 = \frac{DV_{in}}{f_s(\Delta I Li)} \tag{1.2}$$

$$\text{Intermediate capacitor} \quad C_1, C_2 = \frac{DV_{dc}}{[Rf_s(\Delta V_c i)]} \tag{1.3}$$

$$\text{Output filter inductor} \quad L_3, L_4 = \frac{(1 - D)V_{dc}}{[f_s(\Delta I L_O)]} \tag{1.4}$$

$$\text{Output filter inductor} \quad C_0 = \frac{I_{av}}{(2W\Delta V_{dc})} \tag{1.5}$$

From Eqs. (1.1), (1.2), (1.3), (1.4) and (1.5), the components are designed and tabulated as shown in Table 1.1.

1.3 Control Technique

The voltage control mode method is used to regulate the output voltage of the converter. Figure 1.2 displays the block diagram of the circuit using a PI controller.

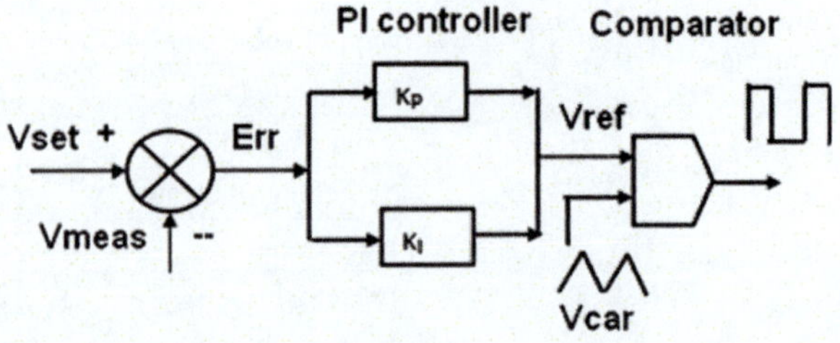

Fig. 1.2 Block diagram of feedback circuit

The comparator gets V_{out} and V_{set} as input. The difference between the two values is called error signal. The error signal generated is given has input to the PI controller. PI controller generates a V_{ref} signal which is compared with the triangle carrier signal. The comparator generates PWM pulses. The nature of the pulse width in the PWM signal varies when we change the error signal. Pulse width of the gating pulse is used to control the V_{out} of proposed converter.

1.4 Simulation Results

The circuit is simulated using MATLAB/Simulink environment. The V_{in} and V_{out} are shown in Fig. 1.4. For V_{in} of 24 V, the V_{out} is stepped up to 48 V and the output is not regulated for any changes in the load side.

1.4.1 Open Loop System

See Figs. 1.3 and 1.4.

1.4.2 Closed-Loop System Using PI Controller

Block diagram of PI controller is shown in Fig. 1.5. Figure 1.6 shows the Simulink closed-loop circuit diagram with PI controller. The function of the comparator in this circuit is, it generates the error which is a comparison between the set value and the measures value [10–12]. This error signal is fed to the proportional and integral block. Proportional and integral block provides reference signal based on the input signal. The function of this reference signal is to regulate the V_{out} by controlling the pulse width of the triggering pulse.

In Fig. 1.7, the input voltage amplitude is 24 V. After $t = 0.35$ s, the supply voltage is increased from 24 to 28 V. The variation given across the supply affects the load voltage. Output voltage is also changed from 48 to 52 V. But due to the PI controller,

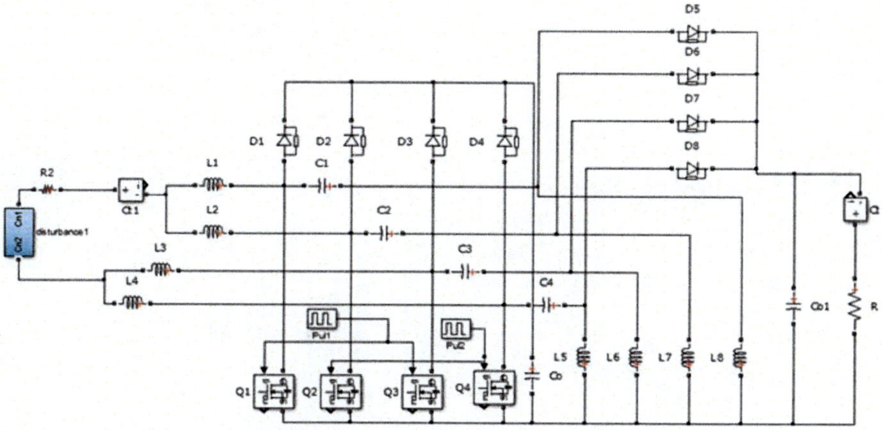

Fig. 1.3 Simulink diagram of bridgeless IBC SEPIC converter in open loop

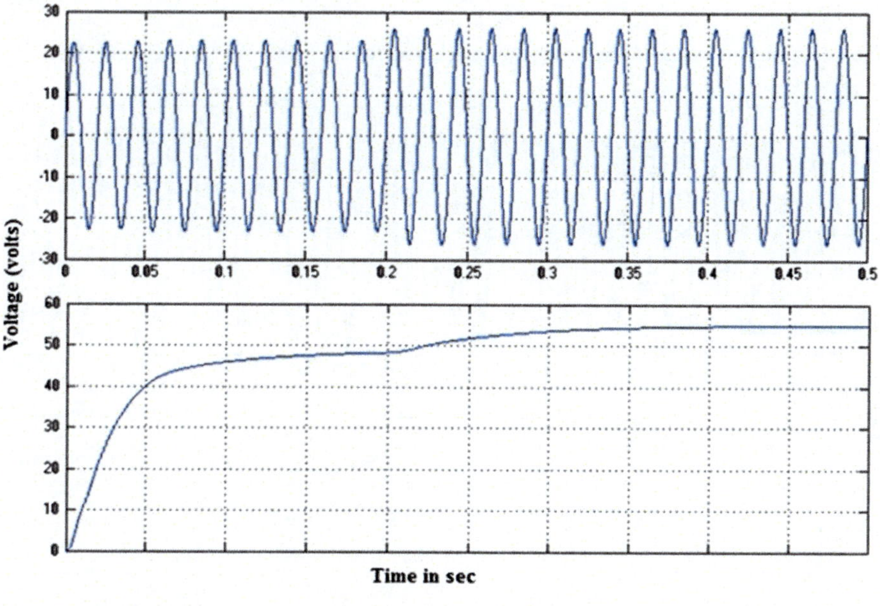

Fig. 1.4 Supply voltage and converter load voltage without regulation

Fig. 1.5 Block diagram of PI controller

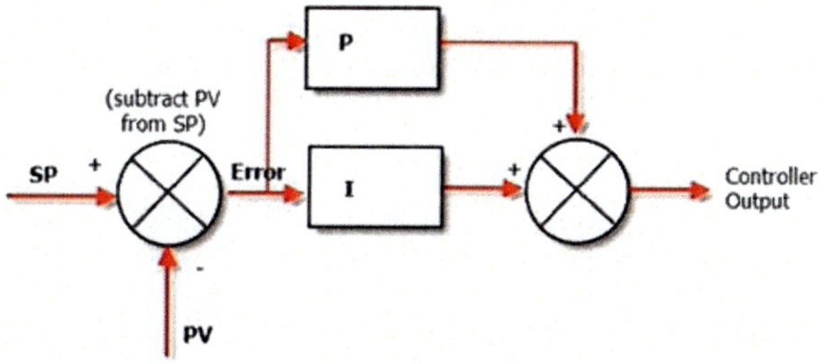

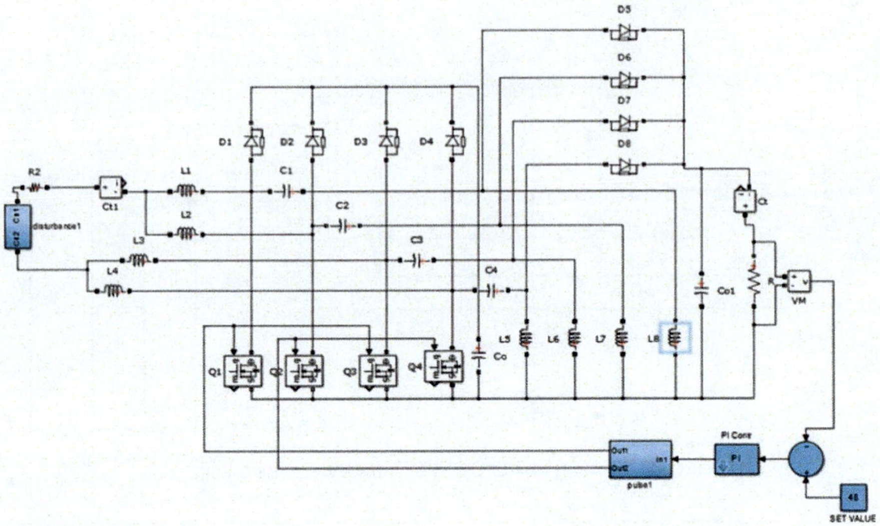

Fig. 1.6 Simulink model of PI controller implemented for bridgeless IBC SEPIC converter

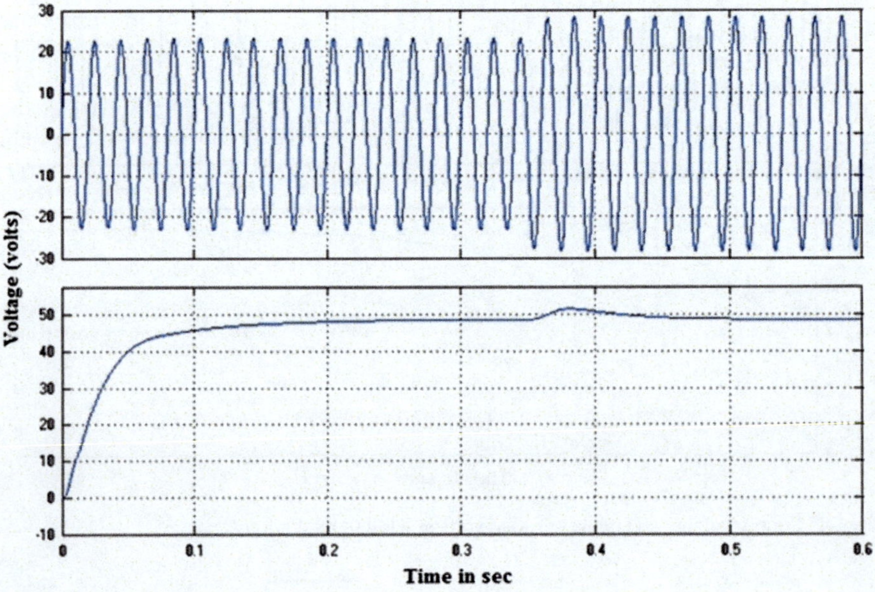

Fig. 1.7 Input voltage and converter load voltage with PI controller

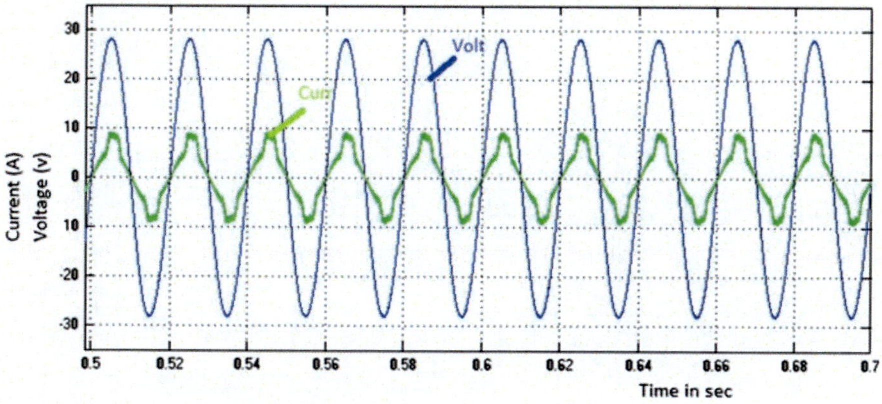

Fig. 1.8 Input voltage and current waveform for converter

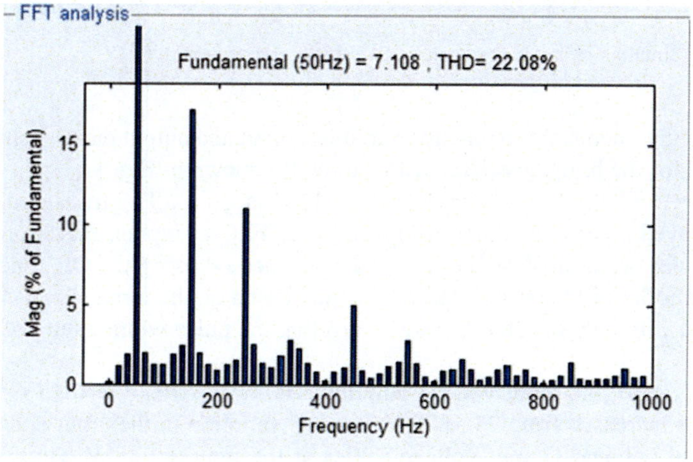

Fig. 1.9 FFT analysis for input current

after $t = 0.4$ s, the output voltage settles at the required 48 V. Thus, the output voltage remains constant when we use closed-loop control method (Fig. 1.7).

1.4.3 Closed-Loop System Using FLC

The simulation diagram for bridgeless IBC SEPIC converter based on FLC is given in Fig. 1.10. For closed-loop operation of FLC, Mamdani fuzzy inference system is preferred for the voltage regulation for the proposed converter. FLC works with

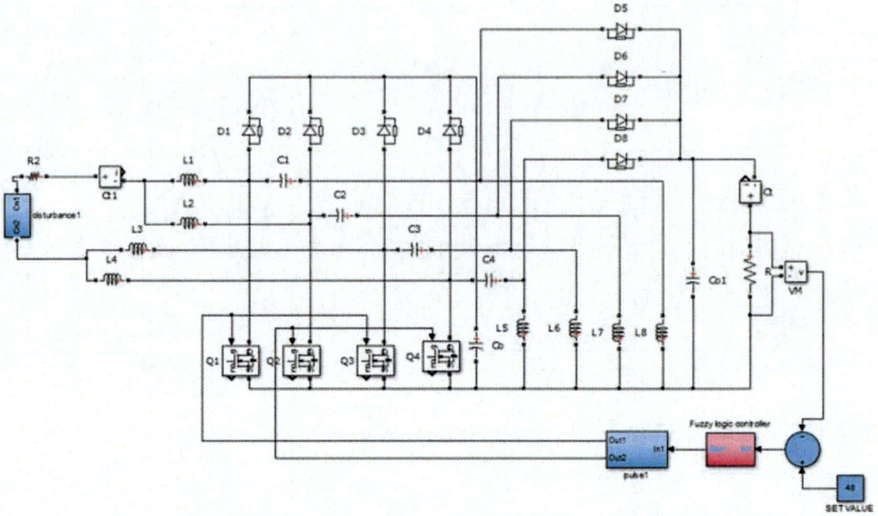

Fig. 1.10 Simulink circuit for bridgeless IBC SEPIC converter with FLC

the triangular membership function variables. Input and output membership function assigned for the bridgeless IBC SEPIC circuit is shown in Fig. 1.11.

In Fig. 1.12 the block diagram of the FLC is given. The measured value and time-delayed measured values are given to the FLC as inputs. Based on this input signal, rules are framed by the trial and error method using the OR functions. The FLC controller produces the reference signal based on the input signal. V_{out} can be regulated with help of reference by controlling the pulse width nature of triggering pulse.

In Fig. 1.13, the input voltage amplitude is 24 V. After $t = 0.15$ s, the supply voltage is increased from 24 to 28 V. But this variation in the input does not affect the output voltage. Output voltage settles at the required 48 V. Thus, the output voltage is maintained constant using a fuzzy logic controller with reduced settling time (Fig. 1.13).

1.5 Results and Discussion

The interleaved SEPIC converter V_{out} is regulated using PI and FLC. The output voltage regulation settling time, overshoot voltage, Steady-state error is computed from the simulation results. Table 1.2 gives the parameter comparison between PI and FLC for the proposed PFC converter. We can observe that FLC results in reduced settling time, peak overshoot, and steady-state error. Hence, the FLC is a suitable one for the PFC converter.

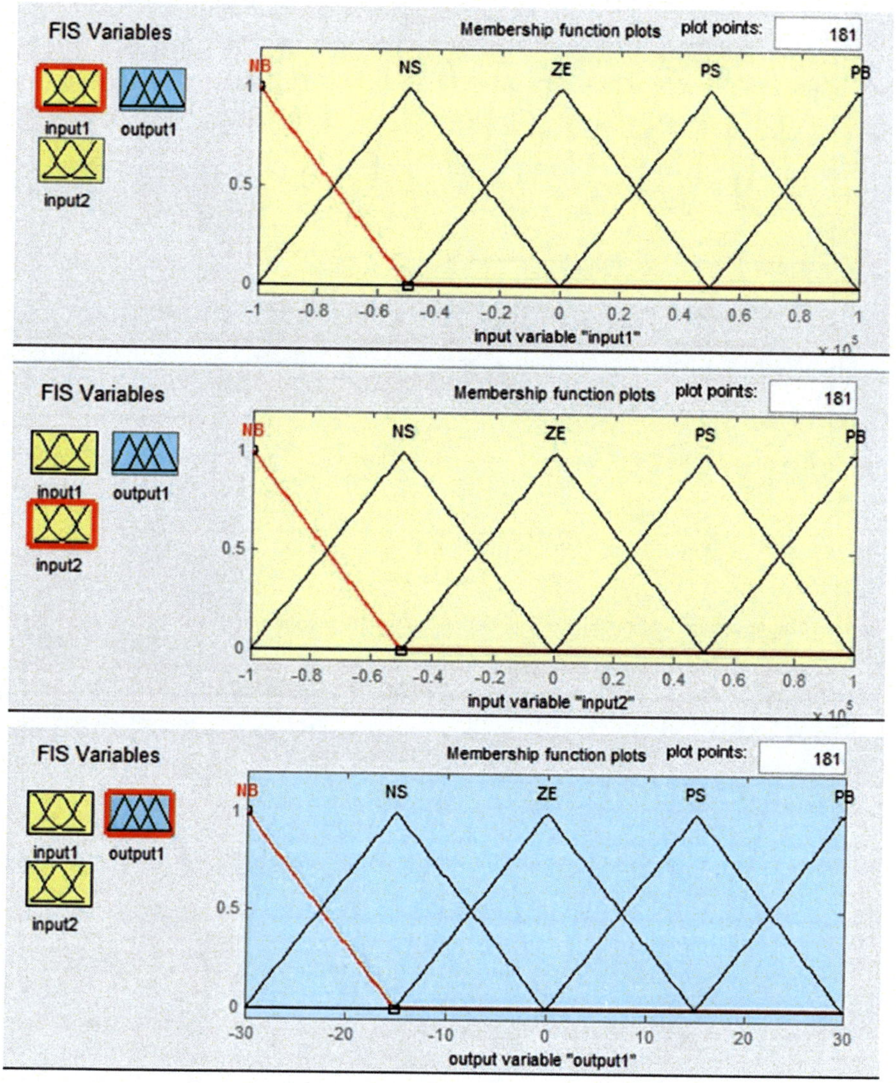

Fig. 1.11 Input and output membership variables

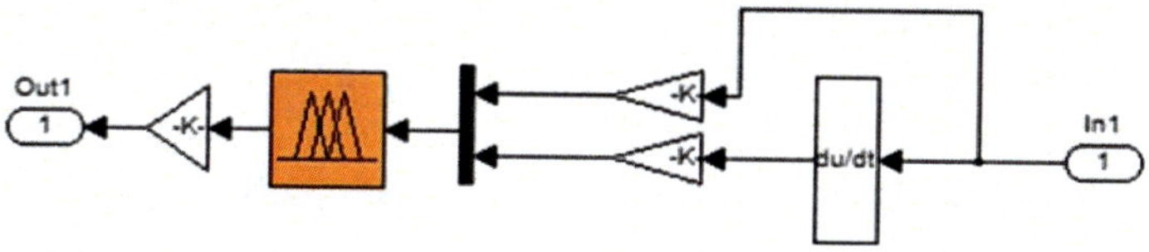

Fig. 1.12 Fuzzy logic controllers

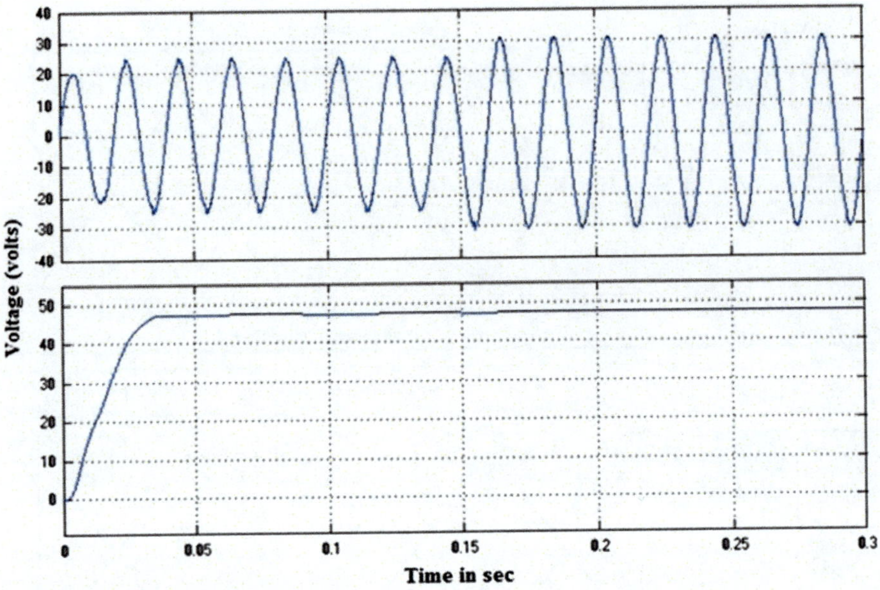

Fig. 1.13 Input voltage and output voltage of converter with regulation

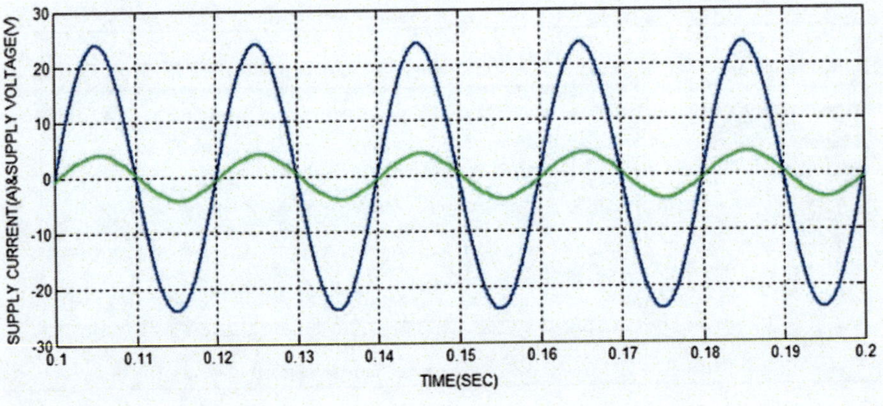

Fig. 1.14 Input voltage and current waveform for converter

1.6 Conclusion

The bridgeless interleaved SEPIC converter fed LED power supply with voltage regulation has been analyzed in this paper. The output voltage is maintained constant using the FLC controller. FLC controller-based converters have an overshoot voltage of 48%. The FLC-based converter gives rise time = 0.001 s, settling time = 0.002 s, steady-state error = 0.2, supply power factor = 0.998, and THD = 15.23%. From

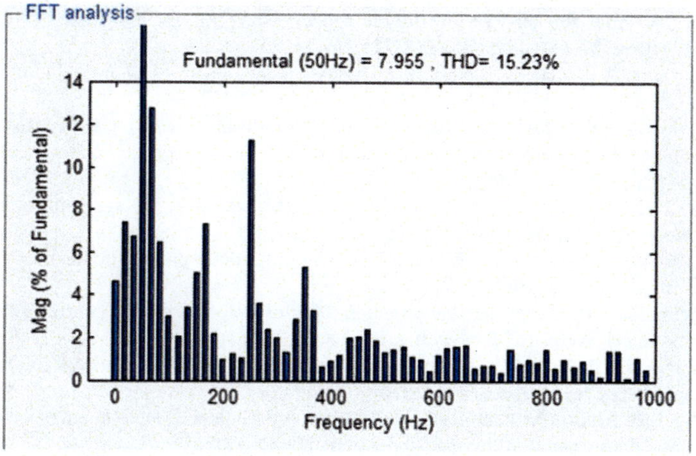

Fig. 1.15 FFT analysis for input current

Table 1.2 Parameter comparison of PI and FLC

Parameter	PI Controller	Fuzzy logic controller
Overshoot voltage should be specified in percentage	52%	48%
Rise time (s)	0.03	0.001
Settling time (s)	0.05	0.002
Steady-state error	0.8	0.2
Supply PF	0.882	0.998
Supply current THD	22.08	15.23

the simulation results, it can be concluded that the FLC-based converter gives better voltage regulation when compared to the PI-based converter.

References

1. E.F. Schubert, *Light-Emitting Diodes* (Cambridge University Press, Cambridge, 2006)
2. J.Y. Tsao, Solid-state lighting: lamps, chips, and materials for tomorrow. IEEE Circ. Devices Mag. **20**(3), 28–37 (2004)
3. C.S. Moo, Y.J. Chen, W.C. Yang, An efficient driver for dimmable LED lighting. IEEE Trans. Power Electron. **27**(11), 4613–4618 (2012)
4. H.J. Chiu, S.J. Cheng, LED backlight driving system for large-scale of LCD panel. IEEE Trans. Ind. Electron. **54**(5), 2751–2760 (2007)
5. X. Long, R. Liao, J. Zhou, Development of street lighting system-based novel high-brightness LED modules. IET Optoelectron. **3**(1), 40–46 (2009)
6. Electromagnetic Compatibility (EMC), Part 3–2: Limits–Limits for Harmonic Current Emissions (Equipment Input Current $\leq$16 A per Phase). International Electro technical Commission IEC 61000-3-2 Standard (2006)

7. S. Wang, K. Yao, Flicker-free electrolytic capacitor-less AC-DC LED driver. IEEE Trans. Power Electron. **27**(11), 4540–4548 (2012)
8. E. Koutroulis, A. Dollas, K. Kalaitzakis, High-frequency pulse width modulation implementation using FPGA and CPLD ICs. J. Syst. Architect. **52**, 332–344 (2006)
9. J. Spangler, A. Behera, A comparison between hysteretic and fixed frequency boost converters used for power factor correction, in *Proceedings of IEEE Applied Power Electronics Conference on Expo* (1993), pp. 281–286
10. S. Park, S. Choi, Soft-switched CCM boost converters with high voltage gain for high-power applications. IEEE Trans. Power Electron. **25**(5) (2010)
11. L. Dixon, Average current mode control of switching mode power supplies, in *Application Note*, Unitrode, U-140 (1999)
12. R. Sreemallika, R. Seyezhai, Investigation of closed-loop current control strategies for bridgeless interleaved SEPIC converter. Int. J. (ELELIJ) **3**(3) (2014)
13. J. Lam, N.A. El-Tawee, A ZV-ZCS electrolytic capacitor-less AC/DC isolated LED driver with continous energy regulation. IEEE (2016)
14. C. Ordiz, J.M. Alonso, M.A. Dalla Costa, J. Ribas, A.J. Calleja, Development of a high-voltage closed-loop power supply for ozone generation. IEEE (2008)
15. G. Udhayakumar, J. BarsanaBanu, Development of a maximum frequency and voltage regulated power supply for ozone generation system, in *International Conference on Recent Trends in Computing, Communication and Networking Technologies (ICRTCCNT'19)*, Kings Engineering College (2019)
16. J.C.W. Lam, P.K. Jain, A TRIAC dimmable single-switch electronic ballast with power factor correction and lamp power regulation. IEEE Trans. Power Electron. **29**(10), 5472–5485 (2014)

Chapter 2
High Gain Converters Based on Coupled Inductors and Gain Extension Cells

Vijay Joseph Samuel, Gna Keerthi, and M. Prabhakar

Abstract DC–DC boost converters with wide voltage conversion ratios are a crucial part of renewable energy systems. In this study, three different converter topologies based on coupled inductor (CI) are discussed. Converter-1 is a conventional boost converter (CBC) with a CI in the place of the simple inductor. In the secondary of the CI, diode capacitor multiplier (DCM) is used to enhance the voltage gain. Converter-2 is an evolution of Converter-1 with additional voltage multiplier cells (VMCs) to improve the voltage gain. Converter-3 is a two-phase interleaved boost converter (IBC) employed with voltage lift technique. A CI is used instead of the simple inductor in one phase of the converter and in order to achieve high voltage gain a DCM cell is connected at the secondary of the CI. Simulation studies are carried out using PSIM to validate the proposed gain extension techniques and design concepts. The proposed topologies are designed and simulated for an input voltage of 24 V DC and an output voltage of 240 V DC while delivering 120 W power.

Keywords DC–DC power converter · Coupled inductor · Diode capacitor multiplier

2.1 Introduction

Power converters with wide voltage conversion ratio play a critical role in interfacing renewable energy systems like photovoltaic (PV) panels. The low voltage available at the output port of the PV panel is often boosted to a more usable voltage value using an intermediate DC–DC converter before supplying the loads.

A CBC cannot be used in high voltage gain applications, as it causes extreme duty ratio operation and diode reverse recovery problems [1]. Hence, an appropriate gain extension mechanism is generally employed to achieve the desired voltage at the load side.

V. J. Samuel · G. Keerthi · M. Prabhakar (✉)
School of Electrical Engineering, Vellore Institute of Technology, Chennai, India
e-mail: prabhakar.m@vit.ac.in

R. Seyezhai et al. (eds.), *Recent Trends in Renewable Energy Sources and Power Conversion*, Springer Proceedings in Energy,
https://doi.org/10.1007/978-981-16-0669-4_2

Various gain extension techniques like voltage multiplier cell (VMC), diode capacitor multiplier (DCM) and voltage doubler are often employed to extend the available voltage gain [2].

Couple inductor (CI)-based gain extension techniques are widely adopted nowadays due to their inherent flexibility in enhancing the voltage gain of the converter by modifying the turns ratio of the CI. Generally, CIs are employed along with several other gain extension methods to achieve the desired output voltage [3].

A CBC using CI with DCM cell in the secondary side of CI is presented in [4]. In [5] and [6], quadratic boost converter (QBC) variants that yield high gain at the output port are discussed. A type of VMC is adopted in [7] to improve the converter's voltage gain. The number of VMCs cells employed determines the voltage gain. A CI-based converter is presented in [8] in which the turns ratio of the CI is adjusted to obtain the desired voltage gain. Hence, the voltage rating of the switch is reduced. Since a simple inductor is used at the input port, continuous current is obtained at the input side.

Often CIs are incorporated into VMCs to further enhance the available voltage gain. Various types of CI and VMC combinations are presented in [9–12]. In [9], a single VM cell with a passive clamp circuit is employed. In [10], a type of VMC using an active clamp circuit is proposed. Furthermore, zero voltage switching is achieved in [10]. A cascaded converter with diode capacitor cell and CI to obtain high gain is proposed in [11]. A symmetrical VM network-based converter is presented in [12].

VMCs are also employed in interleaved multiphase converters. In [13], a two-phase interleaved boost converter (IBC) with VMC and CIs is presented. Similarly, in [14], a two-phase QBC with VMC and CIs is proposed. The converter yields a very high voltage at the output side while achieving low voltage stress on switches. In [15], a combination of DCM cells and CIs is employed to achieve the desired voltage gain.

In this paper, three converters using CI in combination with other gain extension techniques are proposed. The paper is delineated as follows: Sect. 2.1 introduces several recent topologies, Sect. 2.2 describes the circuit evolution of the proposed converters, Sect. 2.3 gives the design details of the converters while simulated results are provided in Sect. 2.4. The proposed converters are compared in Sect. 2.5, and the conclusions are dealt with in Sect. 2.6.

2.2 Circuit Description

2.2.1 Converter-1

In Fig. 2.1a, the power circuit diagram of Converter-1 is presented. The converter consists of a CBC, to which a diode capacitor multiplier (DCM) cell is integrated in the secondary side of the CI to achieve higher voltage gain.

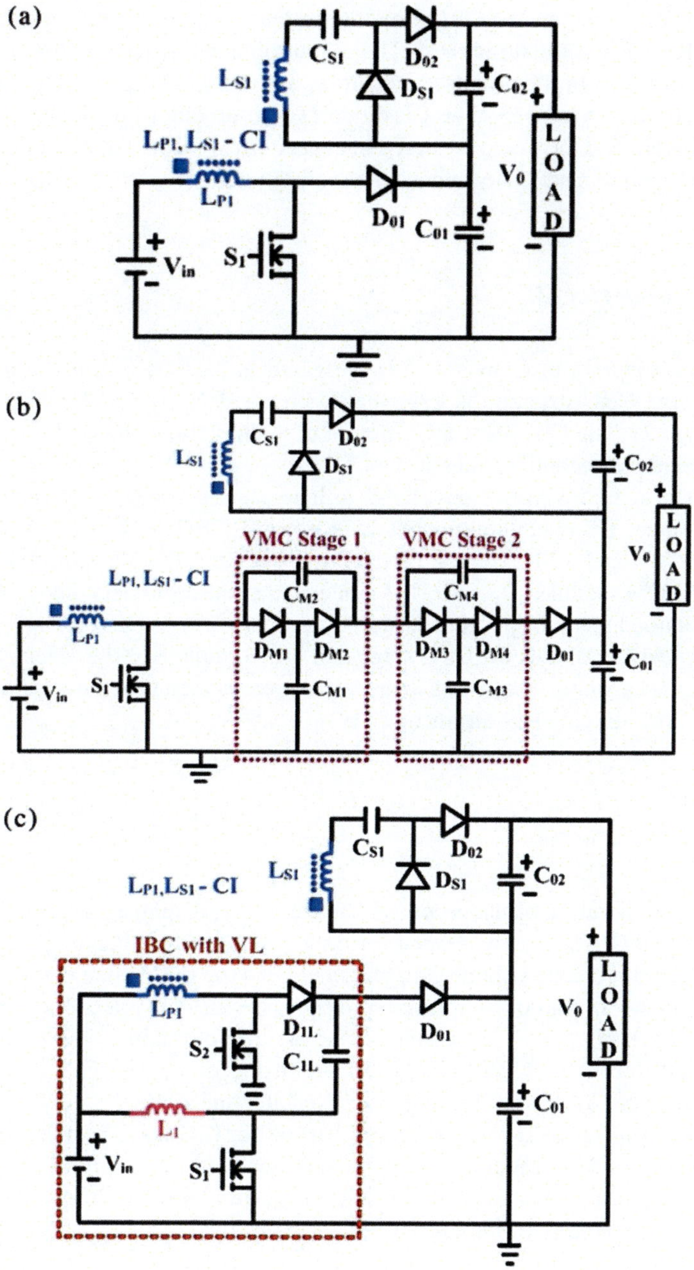

Fig. 2.1 Power circuit configuration of **a** Converter-1, **b** Converter-2, **c** Converter-3

A CI is employed in the CBC structure instead of a simple inductor to increase the gain by varying the turns ratio. The output obtained from the CBC is stored in the capacitor C_{01}. In the proposed converter, the DCM cell ($D_{S_1} - C_{S_1}$) is attached to the secondary winding of the CI, diode D_{02} is the output rectifier diode for the multiplier cell, and the output is stored in capacitor C_{02}. Both the capacitors C_{01} and C_{02} are cascaded in series to achieve the required voltage gain at the load side.

2.2.2 Converter-2

The circuit diagram of Converter-2 is presented in Fig. 2.1b. In this topology, the output of the CBC structure is integrated with two VMC units (VMC stage 1 and VMC stage 2). The CBC uses a CI instead of a simple inductor and a DCM cell is embedded at the secondary side of the CI.

VMCs are employed in Converter-2 to increase the voltage gain obtained from the CBC stage. The overall output obtained from the CBC, VMC stage 1 (D_{M_1}, D_{M_2}, C_{M_1} and C_{M_2}) and VMC stage 2 (D_{M_3}, D_{M_4}, C_{M_3}, and C_{M_4}) is stored in capacitor C_{01}. The DCM module ($D_{S_1} - C_{S_1}$) is connected at the secondary side of CI and the output obtained from this module is stored in capacitor C_{02}. Diodes D_{01} and D_{02} act as output rectifier diodes for CBC stage and DCM module, respectively. Similar to Converter-1, capacitors C_{01} and C_{02} are serially connected to achieve higher voltage gain and fulfill the load requirement.

2.2.3 Converter-3

The circuit illustration of the proposed Converter-3 is displayed in Fig. 2.1c. In this structure, two CBCs are interleaved forming an IBC, also voltage lift technique is employed to double the voltage gain obtained from the IBC which comprises of one simple inductor L_1 and one CI. Similar to the previous converters (Converter-1 and 2), a DCM cell ($D_{S_1} - C_{S_1}$) is connected across the secondary winding of the CI.

In Converter-3, the CBCs are interleaved using diode D_{1L} and capacitor C_{1L}, where capacitor C_{1L} acts as the voltage lift capacitor and diode D_{1L} is the intermediate rectifier diode. The voltage gain obtained from the IBC stage is transferred to the output capacitor C_{01} through the output rectifier diode D_{01}. The output obtained from the DCM cell is stored in the capacitor C_{02}. Similar to above stated converters, capacitors C_{01} and C_{02} are cascaded in series to meet the output voltage requirement. The complete operating principle of all the three converters is based on the working of individual converters and gain extension mechanisms discussed in [4] and [15].

Table 2.1 Voltage conversion ratio

	Converter-1	Converter-2	Converter-3
Voltage across C_{01}	$\frac{1}{1-D}V_{in}$	$\frac{3}{1-D}V_{in}$	$\frac{2}{1-D}V_{in}$
Voltage gain	$M_1 = \frac{1+nk}{1-D}$	$M_2 = \frac{3+nk}{1-D}$	$M_3 = \frac{2+nk}{1-D}$

2.3 Analysis During Steady-State Condition and Design Details

The steady-state voltage gain, semiconductor stress, and design details of inductors and capacitors are given in this section.

2.3.1 *Voltage Conversion Ratio*

The voltage gain of the three converters is obtained using volt-second balance. The net output in three converters is the addition of voltages across C_{01} and C_{02}. Since a DCM module is used in the secondary side of CI in the three converters, the voltage across C_{02} is given by,

$$C_{02} = \frac{nk}{1-D}V_{in} \tag{2.1}$$

where D is the duty ratio of the switch, n is the turns ratio of CI, k is the coupling co-efficient, and V_{in} is the input voltage.

The voltage developed across the capacitor C_{01} differs in the three capacitors due to the different voltage extension mechanism employed. In Converter-1, a CBC is used; hence, the voltage across C_{01} is same as the output available in a CBC.

Since 2 VMC cells are used in Converter-2, the voltage built across C_{01} is thrice the output voltage of a CBC. In Converter-3, due to voltage lift technique, the voltage across C_{01} is twice the output of a conventional boost. The voltage developed across C_{01} and the voltage gain of all the three converters is given in Table 2.1.

2.3.2 *Switch Stress*

Voltage stress on switch

Since a CBC structure is employed in all the proposed topologies, the voltage stress experienced by respective CBC switches is equivalent to the CBC output, and voltage rating of the switch is computed by equation mentioned in Table 2.2.

Table 2.2 Stress on the semiconductors in the converters

	Switch voltage	Switch current	Diode voltage
Converter-1	$V_{sw} = \frac{1}{1-D} V_{in}$	I_{in}	$V_{D01} = \frac{1}{1-D} V_{in}$, $V_{D02}, V_{DS_1} = \frac{nk}{1-D} V_{in}$
Converter-2		I_{in}	$V_{D02}, V_{DS_1} = \frac{nk}{1-D} V_{in}$, All other diodes $\frac{1}{1-D} V_{in}$
Converter-3		$I_{S_1} = I_{in}, I_{S_2} = \frac{I_{in}}{2}$	$V_{D1L}, V_{D01} = \frac{1}{1-D} V_{in}$, $V_{D02}, V_{DS_1} = \frac{nk}{1-D} V_{in}$

Current stress on switch

In converters 1 and 2, the switches are located near the input port of the CBC structure; hence, the current stress impresses across the switches is equivalent to input current and can be calculated by the equation stated in Table 2.2.

In Converter-3, an IBC structure is employed; hence, current stress experienced by switch S_2 is equivalent to half of input current, whereas switch S_1 is exposed to higher value of current stress due the employment of voltage lift technique. Current stress experienced by both S_1 and S_2 can be determined by the equation presented in Table 2.2.

2.3.3 Inductor and Capacitor Design

The equation for designing the inductance and capacitance values is given as

$$L = \frac{V_{in} \times D}{f \times \Delta i_L} \text{ and } C = \frac{I_C \times D}{f \times \Delta v_C} \tag{2.2}$$

The inductance value of secondary winding of the CI is given by

$$L_{S_1} = n^2 \times L_{P_1} \tag{2.3}$$

where V_{in} is the supply voltage, D refers to the duty ratio, f refers to the switching frequency, ΔiL and Δv_C respectively indicate the inductor ripple current and ripple voltage across capacitor, and I_C refers to the capacitor current.

2.4 Discussion on Simulation Results

To validate the proposed concepts, the converters are simulated in PSIM environment. The converters are designed and simulated for an input voltage of 24 V and output voltage of 240 V while delivering 120 W power. The turns ratio of all the converters is selected as $n = 3$. The duty ratio is calculated for each converter to realize a voltage gain of 10.

2.4.1 *Converter-1*

Voltage gain capability of Converter-1 is confirmed through Fig. 2.2a. When 24 V DC input is applied, DC output of 240 V is obtained at the load end as desired. A 50 kHz gating signal with 0.62 duty ratio is given to the gate of the switch S_1. An output current of 0.5 A validates the output power of 120 W.

Figure 2.2b presents the voltage stress on the switches and diodes. The voltage stress impressed on S_1 is only about one-fourth (25%) of the output voltage. Among the diodes, voltage stress on D_{01} is only 60 V, while the voltage stress on D_{S_1} and D_{02} is about 180 V. The simulated values agree with their theoretical counterparts.

The input and inductor currents are shown in Fig. 2.2c. The currents flowing through the primary and secondary windings of the CI are linear and their values are as expected. The switch current follows the inductor current when the switch is in ON state.

2.4.2 *Converter-2*

Figure 2.3a validates the voltage gain ability of Converter-2. For a 24 V input, 240 V output at 120 W is obtained with a duty ratio of 0.44. Figure 2.3b, c depicts the voltage stress on the semiconductor devices. Voltage stress on S_1 is only 16.66% of output voltage.

The voltage stress impressed across all the diodes is in concurrence with the equations presented in Table 2.1. The operation of the diodes in a VMC cell is complementary which can be verified from the figure. Voltage across the diodes D_{M_1} and D_{M_2} is complementary and similar to the voltage across D_{M_3} and D_{M_4}. The currents flowing through the individual primary and secondary windings of the CI are similar to that of Converter-1 which can be verified from Fig. 2.3d. The current flowing through diode D_{02} follows the inductor current L_{S_1} as observed from the figure.

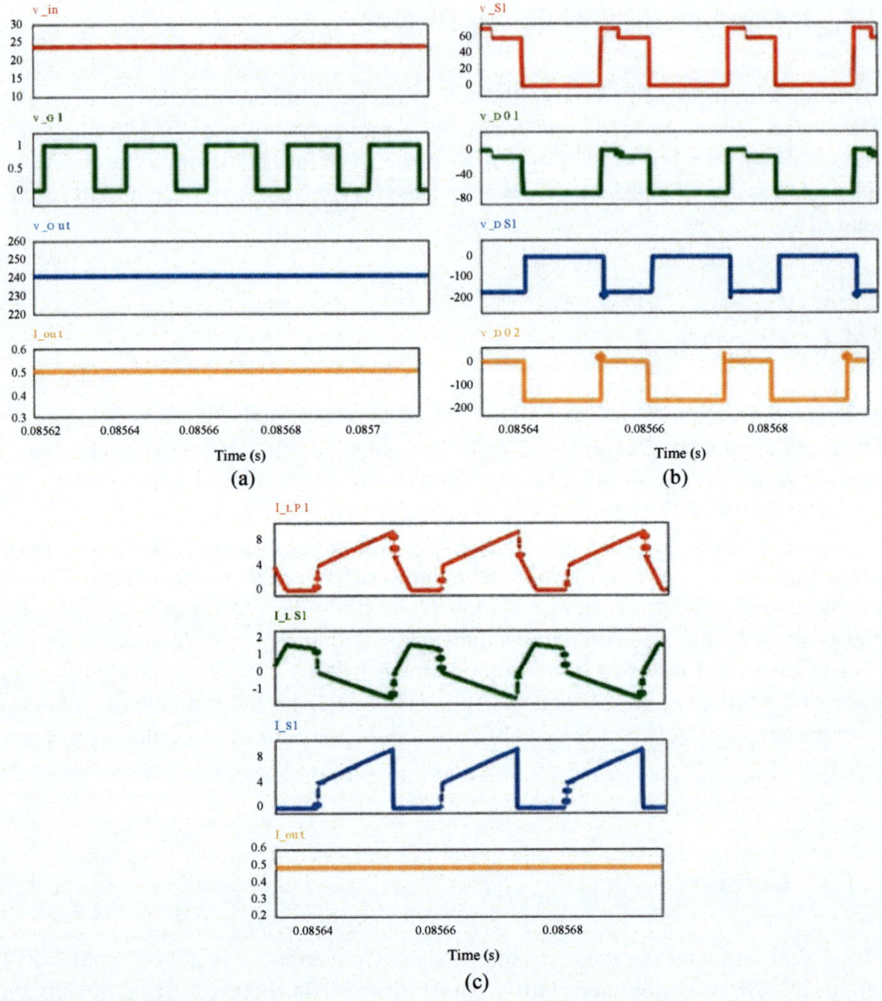

Fig. 2.2 Simulation results of Converter-1. **a** Input voltage, gate pulse to S_1, output voltage, output current. **b** Voltage stress on S_1, D_{01}, D_{S_1}, D_{02}. **c** Current flowing through L_{P_1}, L_{S_1}, S_1 and output current

2.4.3 Converter-3

Converter-3 is a two-switch variant unlike the other two. The two switches are operated with a 180° phase shift at 0.52 duty ratio. The converter yields an output of 240 V when 24 V is supplied at the input. These can be verified from Fig. 2.4a. The stress on the switches is shown in Fig. 2.4b. The voltage stress on the switches is only 20% of V_0 which can be seen from the image. The simulated values are in agreement with their theoretical values. The voltage stress on diodes is shown in Fig. 2.4c. The

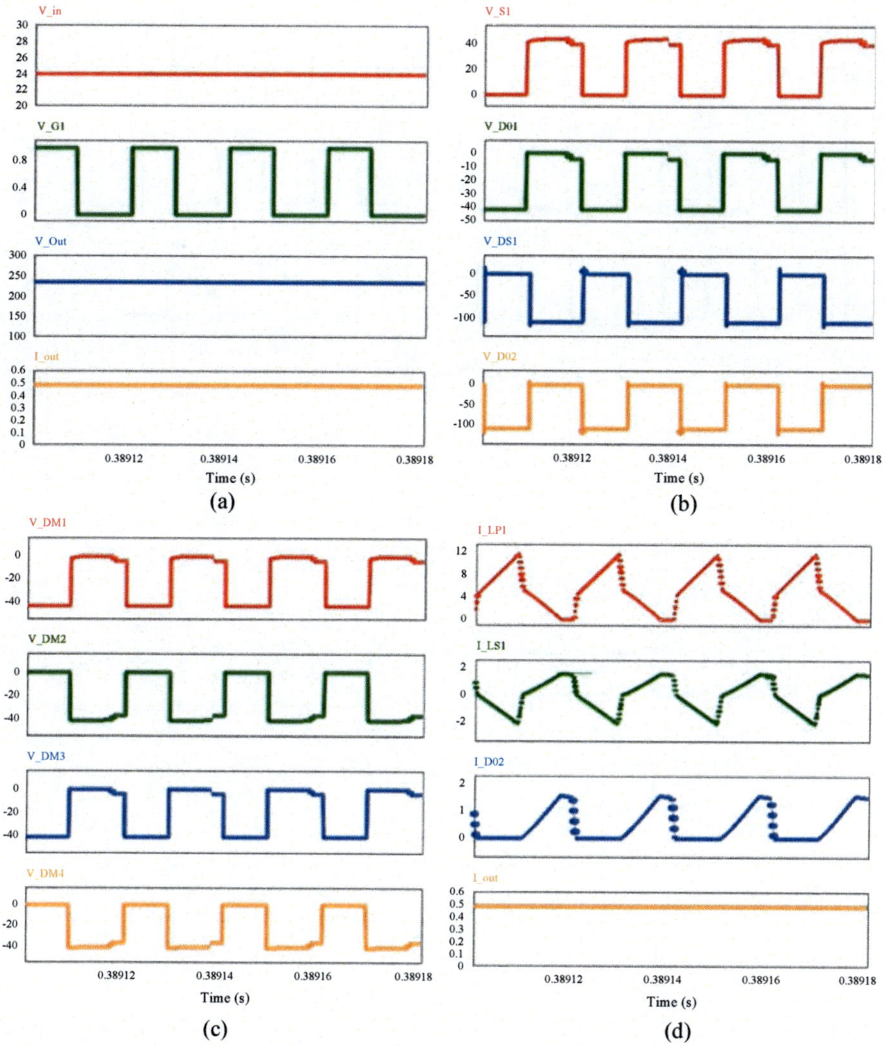

Fig. 2.3 Simulation results of Converter-2. **a** Input voltage, gate pulse to S_1, output voltage and output current. **b** Voltage stress on S_1, S_2, D_{01}, D_{S_1}, and D_{02}. **c** Voltage stress on D_{M_1}, D_{M_2}, D_{M_3}, and D_{M_4}. **d** Currents flowing through L_{P_1}, L_{S_1}, D_{02}, and output current

stress value of diode D_{01} is the least value which is only 20% of the output voltage. The stresses on other diodes match with the values predicted using equations in Table 2.1.

The current waveforms are shown in Fig. 2.4d. Due to the transformer action of the CI, the input current is not continuous and its pulsed. The linear charging and discharging action of L_1 is clearly observed. The current waveforms of L_{P_1} and L_{S_1} are similar to the other two converters due to their structural similarity.

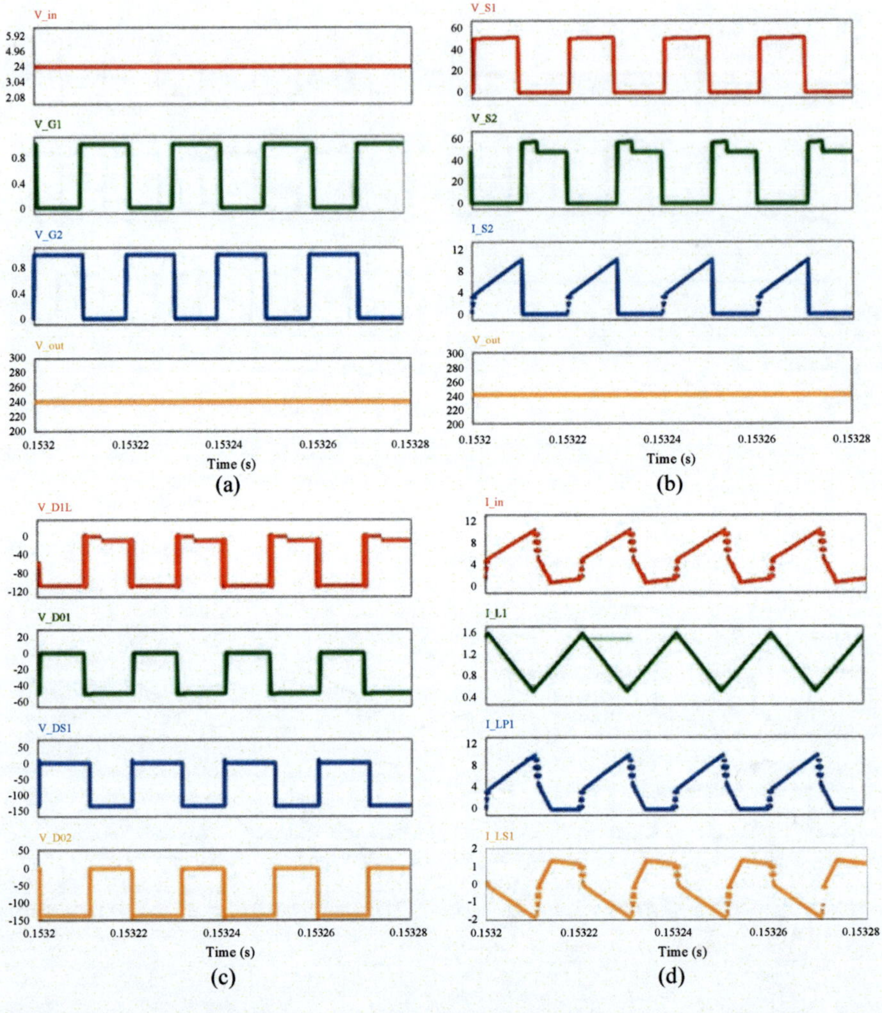

Fig. 2.4 Simulation results of Converter-3. **a** Input voltage, gate pulse to S_1, gate pulse to S_2, output voltage. **b** Voltage stress on S_1, S_2, current stress on S_2 and output voltage. **c** Voltage stress on D_{1L}, D_{01}, D_{S_1}, D_{02}. **d** Current flowing through input, L_1, L_{P_1}, and L_{S_1}

2.5 Comparison of the Proposed Converters

The voltage gain plot is shown in Fig. 2.5. The voltage gain of Converter-2 is the highest of all the three converters while Converter-1 operates with the least voltage gain value. However, Converter-2 uses the highest number of components (17) while Converter-1 uses only 8. Though a gain of 10 is achieved in all the three converters, they all operate at different duty ratios; the CIs are designed with the same turns ratio. The voltage stress on the switch used in Converter-2 is only 16.66% of V_0.

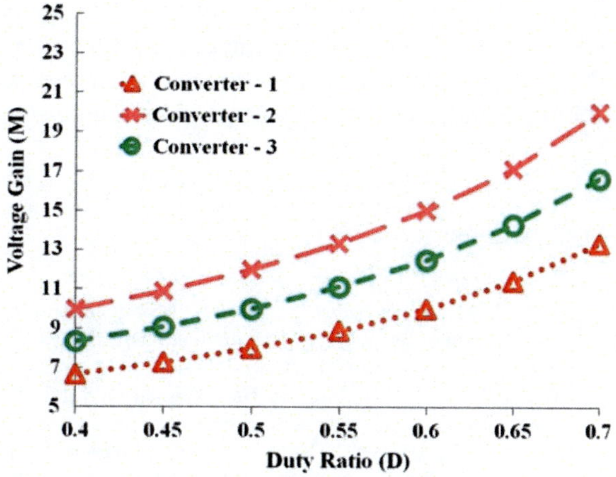

Fig. 2.5 Voltage gain plot of the proposed converters with $n = 3$

Table 2.3 Comparison of the proposed converters

Attributes	Converter-1	Converter-2	Converter-3
Duty ratio	0.62	0.44	0.52
Magnetic elements	1 (multi-winding CI)	1 (multi-winding CI)	2 (1 multi-winding CI, 1 simple inductor)
Turns ratio	3	3	3
No. of switches	1	1	2
No. of diodes	3	8	4
Voltage stress on switches as percent-age of V_O (V_{sw})	25	16.66	20
Total component count (TCC)	8	17	12
Gain extension mechanism	CI + DCM	CI + DCM + VMC	CI + DCM + voltage lift

The voltage stress on the switches in Converter-1 and 3 is 25% and 20% of V_0 respectively. Important parameters are compared in Table 2.3.

2.6 Conclusion

Three CI-based converters with wide voltage conversion ratios were proposed and compared. Along with CIs, different voltage gain extension techniques like VMC, voltage lift, and DCM were adopted in each converter to realize the required voltage gain of 10. The switches employed in all the three converters were subjected to low voltage stress values only. Simulation results obtained from PSIM validate the adopted gain extension techniques and design concepts. The converters were designed and simulated for an input voltage of 24 V DC, output voltage of 240 V DC at 120 W power. Based on the simulation results, the three converters are good choice for integrating a low voltage PV to a high voltage bus.

References

1. T.R. Choudhury, B. Nayak, A. De et al., A comprehensive review and feasibility study of DC–DC converters for different PV applications: ESS, future residential purpose. EV Charging Energy Syst. **11**, 641–671 (2020)
2. B. Sri Revathi, M. Prabhakar, Non isolated high gain DC-DC converter topologies for PV applications—a comprehensive review. Renew. Sustain. Energy Rev. **66**, 920–933 (2016)
3. H. Liu, H. Hu, H. Wu, Y. Xing, I. Batarseh, Overview of high-step-up coupled-inductor boost converters. IEEE J. Emerg. Sel. Topics Power Electron. **4**(2), 689–704 (2016)
4. H.-W. Seong, H.-S. Kim, K.-B. Park, G.-W. Moon, M.-J. Youn, High step-up DC-DC converters using zero-voltage switching boost integration technique and light-load frequency modulation control. IEEE Trans. Power Electron. **27**(3), 1383–1400 (2012)
5. V.J. Samuel, G. Keerthi, M. Prabhakar, High gain interleaved quadratic boost DC-DC converter, in *2019 2nd International Conference on Power and Embedded Drive Control (ICPEDC)*. IEEE (2019)
6. V.J. Samuel, G. Keerthi, P. Mahalingam, Interleaved quadratic boost DC–DC converter with high voltage gain capability. Electr. Eng. **102**, 651–662 (2020)
7. B. Zhu, H. Wang, D.M. Vilathgamuwa, Single-switch high step-up boost converter based on a novel voltage multiplier. IET Power Electron. **12**(14), 3732–3738 (2019)
8. Y. Zheng, K.M. Smedley, Analysis and design of a single-switch high step-up coupled-inductor boost converter. IEEE Trans. Power Electron. **35**(1), 535–545 (2020)
9. M. Premkumar, C. Kumar, A. Anbarasan et al., A novel non-isolated high step-up DC–DC boost converter using single switch for renewable energy systems. Electr. Eng. **102**, 811–829 (2020)
10. A. Mirzaee, J.S. Moghani, Coupled inductor-based high voltage gain DC–DC converter for renewable energy applications. IEEE Trans. Power Electron. **35**(7), 7045–7057 (2020)
11. Q. Pan, H. Liu, P. Wheeler, F. Wu, High step-up cascaded DC–DC converter integrating coupled inductor and passive snubber. IET Power Electron. **12**(9), 2414–2423 (2019)
12. W. Hassan, D.D.-C. Lu, W. Xiao, Single-switch high step-up DC–DC converter with low and steady switch voltage stress. IEEE Trans. Ind. Electron. **66**(12), 9326–9338 (2019)
13. Y. Zheng, W. Xie, K.M. Smedley, Interleaved high step-up converter with coupled inductors. IEEE Trans. Power Electron. **34**(7), 6478–6488 (2019)
14. V.J. Samuel, G. Keerthi, P. Mahalingam, Coupled inductor-based DC–DC converter with high voltage conversion ratio and smooth input current. IET Power Electron. **13**(4), 733–743 (2020)
15. B. Sri Revathi, P. Mahalingam, Non-isolated high gain DC–DC converter with low device stress and input current ripple. IET Power Electron. **11**(15), 2553–2562 (2018)

Chapter 3
Stability Characteristics of Photovoltaic System Using SVC Controllers

P. Aruna Jeyanthy, S. Raja Mohamed, D. Devaraj, J. D. Darwin, and Mohammed Alarfaj

Abstract In the present era, solar photovoltaic of renewable power generation has helped human mankind owing to the depletion of fossil fuels for the power generation. Due to the intermittent nature of solar power, many problems are faced by operators during the power generation. This research work proposes a new methodology to control the operating behaviour by including the static VAR compensator (SVC), to damp the unnecessary fluctuations and to enhance the dynamic behaviour of the solar photovoltaic power generation, under various levels of PV penetrations. The critical clearing time, rotor swing angle and rotor deviation parameters are determined to know the dynamic characteristics. The improvement of the stability validates the work. A comparison study for the stability with and without the SVC is also presented here.

Keywords Photovoltaic (PV) · Static VAR compensator (SVC) · Transient stability · Critical clearing time · Modelling

3.1 Introduction

The shortfall in power generation is overcome due to the lack of power produced by traditional energy sources, as well as the pollution-free generation of limitless renewable energy [1]. Solar PV generation is common and easily affordable to all. The worldwide solar capacity has gone up to the rate of 56% [2]. The isolated and the grid tied are the two types of the solar power generation. The grid tied solar power system is only helpful to connect the entire centralized power network. But, the challenges faced by the operators due to the link between them are enormous.

P. Aruna Jeyanthy (✉) · D. Devaraj
Kalasalingam Academy of Research and Education, Srivilliputhur, India

S. Raja Mohamed · M. Alarfaj
Department of Electrical Engineering, King Faisal University, Al Hofuf, Saudi Arabia

J. D. Darwin
Loyola Institute of Technology and Science, Thovalai, India

R. Seyezhai et al. (eds.), *Recent Trends in Renewable Energy Sources and Power Conversion*, Springer Proceedings in Energy,
https://doi.org/10.1007/978-981-16-0669-4_3

The few problems encountered are the inertial changes and synchronization to the grid lines. If these problems are not rectified, the fluctuation arises and collapses the operating system. Hence, the impact of solar PV due to the fluctuations has predominant problems for bulk power generation [3]. But these problems are to be solved simultaneously, since the whole world nowadays aims to rely on solar power due to its added benefits such as clean and abundant freely. The design of suitable inverter/controller and active/reactive controller has partially solved the problems. Recent research has solved most of the issues, still when high solar PV power is allowed to penetrate the grid, it reduces the required inertia and affects the frequency stability [4]. Authors in [5] studied the behaviour aspect during the above situations using IEEE-39 casework. The dynamic performance of the Ontario network during high penetration of the solar PV is analysed and discussed by the authors in [6]. Terrace covered with layers of solar cells connected to the grid is analysed in [7]. Different rate of flow of penetration of solar power was studied, and it was concluded that 20% of them can perform well even in the fault conditions. Its influence in the single machine infinite bus system in the transient state was discussed in [8]. Wind farms and hybrid renewable energy systems with integrated power systems will benefit from FACTS controllers [9, 10]. The dynamic state of the network is also better in [11] for the two areas- an integrated device using ANFIS-PID controller, even when faults occur. The simulation results show better robustness and stability compared to the conventional PID controller.

In this proposed work, the improvement in the dynamic performance is evaluated in a PV farm consisting of two machines running simultaneously with a power system stabilizer (PSS) and an SVC controller under various penetration levels. At first, multi-system modelling was developed and elaborated in Sect. 3.2. Then, a dynamic-aggregated PV farm model with the PSS and the SVC controllers are modelled using MATLAB Simulink software. Section 3.3 covers the modelling of the solar PV farm and SVC controller. The problems associated with the dynamic performance for the entire work are represented in Sect. 3.4. Modelling and analysis of the proposed work and the Simulink results are given in Sect. 3.5.

3.2 Model System

3.2.1 Power System Model

The two area machine called the multi-area machine is modelled for the proposed work. The MATLAB test model is shown in Fig. 3.1a. The generating capacity of the two machines is 1000 MW. The connecting link between the loads and the machines has the capacity of 500 kV and 700 km length. 5000 MW resistive load is in the receiving side of the network. The generators are activated with the hydraulic model, consisting of the PSS for stabilizing the operating constraints. In the middle

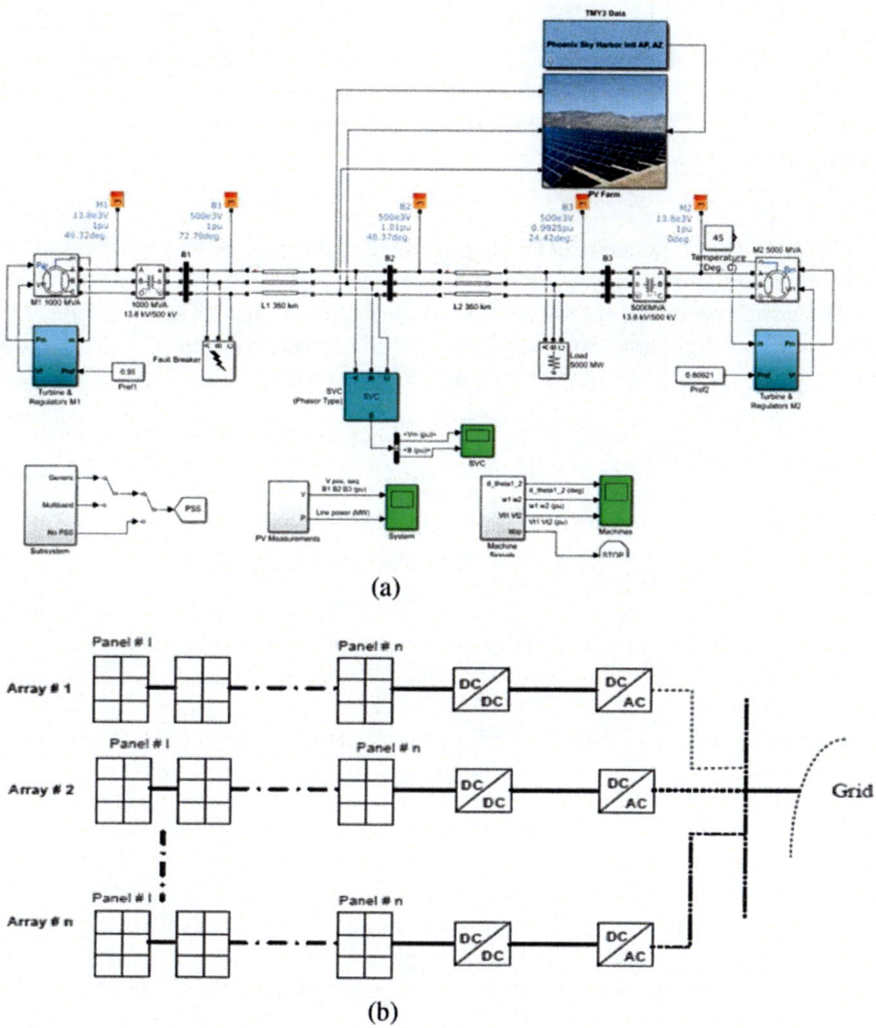

Fig. 3.1 **a** MATLAB test model, **b** schematic diagram of PV systems with N arrays [12]

of the system, the solar PV farm is connected, and the high penetrations for various percentages of values are evaluated without SVC and with SVC controllers.

3.2.2 Solar PV Farm Model

An ordered structure of the PV panel is modelled for the solar PV farm to form an array. Then, they are interfaced between the power electronic converters as shown

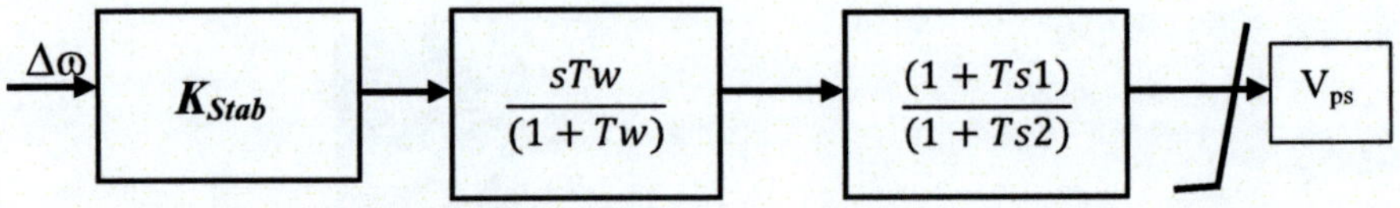

Fig. 3.2 Block diagram of PSS

in Fig. 3.1b. The complete structure is connected to the gridline through the point of common coupling.

The power generated by the solar PV farm is the product of the percentage of efficiency and the region enclosed by it with the irradiation effect. It is given in Eq. (3.1). The percentage of penetration level is shown in Eq. (3.2).

$$P_{\text{gen}} = \%\eta \times \text{Area}_{\text{PV}_{\text{farm}}} \times \text{Irradiance (in w/m}^2\text{)} \tag{3.1}$$

PV penetration is defined as

$$\%\text{PV penetration} = \frac{\text{Total PV Generation (MW)}}{\text{Total generation (MW)}} \tag{3.2}$$

3.3 Modelling of PSS and SVC Controller

3.3.1 PSS Model

The PSS generates the twisting moment of force in the axis of the rotor generator and provides compensation due to the displacement that exists in the operation and improves the stability. The block diagram is given in Fig. 3.2. The differential angular speed signal is fed to the stabilizing block, which is cascaded to the high pass filter and the displacement network.

3.3.2 SVC Model

SVC contribution in this proposed work is to absorb/generate reactive power and hence to control the voltage [13].

The operating current equation is given as

$$i_{\text{TCR}} = V_m B_{L(\alpha)} \tag{3.3}$$

Fig. 3.3 Block diagram of SVC

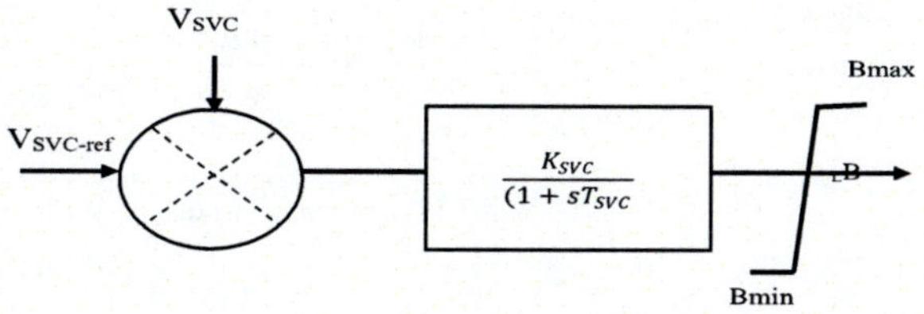

where

V_m voltage in the SVC nodal point.

$$B_{L(\alpha)} = \frac{2(\pi - \alpha) + \sin(2\alpha)}{\pi X_L}; \quad \pi/2 \leq \alpha \leq \pi \tag{3.4}$$

Susceptance of the SVC modelling is given as

$$B_{\text{SVC}}(\alpha) = \frac{1}{XL} - B_L(\alpha) \tag{3.5}$$

$XL = L\omega$ is the reactance of the reactor

$$B_L = \frac{1}{T_{\text{SVC}}}(-B_L(t) + B_{\text{LO}} + K_{\text{SVC}} U_B(t)) \tag{3.6}$$

where

$B_L(t)$ represents the inductance susceptance
B_{LO} represents the susceptance in the initial state
T_{svc} represents the time constant
K_{svc} represents the SVC gain regulating system
$U_B(t)$ represents the input of the SVC regulator.

The block diagram representation of SVC is shown in Fig. 3.3. The major role of static VAR compensator is adjusting the voltage at its terminals [14, 15].

3.4 Impact of Power System Dynamic Stability and Discussions

In general, the fluctuations are caused due to the failure of any devices, line and node faults, natural calamities and mishandling of the operations. The phenomenon of being in synchronism during this time is known as the dynamic stability characteristics. Simulation using MATLAB/Simulink software is performed below. The simulation parameters are shown in Appendix. The proposed work observes the

Table 3.1 Study scenarios

Type of disturbance	System model	PV capacity (MW)[a]
Single-phase fault	Without controller	19
	With controller	19
Three-phase fault	Without controller	19
	With controller (10% penetration)	9.5
	With controller (20% penetration)	19
	With controller (40% penetration)	38
	With controller (50% penetration)	47.5

[a]PV capacity (MW) is calculated using Eq. (3.1)

different fluctuations existing during the high solar PV penetrations in the grid. The short circuit faults are considered for the time duration of 5 s, with the closure of fault after 0.1 s. The total running period is 50 s.

3.4.1 Scenario Description

Table 3.1 shows the investigation of the performance of PSS-SVC controller on dynamic stability.

Case-I. Single-Phase Fault

Without the controller, the maximum speed deviation is 0.00031 (PU), oscillation duration is 3.97 s, and it satisfies the standard condition (ref: Fig. 3.4). In this case, particular rotor speed is stable. At the time of the fault, the rotor angle difference between two machines reached a limit of 92°, which is less than the unstable case rotor angle (120°). As a result, the system in this case is stable.

Figure 3.5 indicates the change in the rotor angle, rotor variation and terminal voltages. From the rotor angle plot, clearly, it is observed that the fluctuations are largely reduced without the controller as in case (i). The statistical measurement of the signal has the maximum speed deviation of 0.00026 (PU), oscillation duration is 3.52 s, and it satisfies the standard condition for the stable range of rotor speed deviation at the particular time interval. In this case, the particular rotor speed is stable. At the time of the fault, the rotor angle difference between two machines reached a limit of 64.26°, which is less than the unstable case rotor angle (120°). As a result, the system in this case is stable.

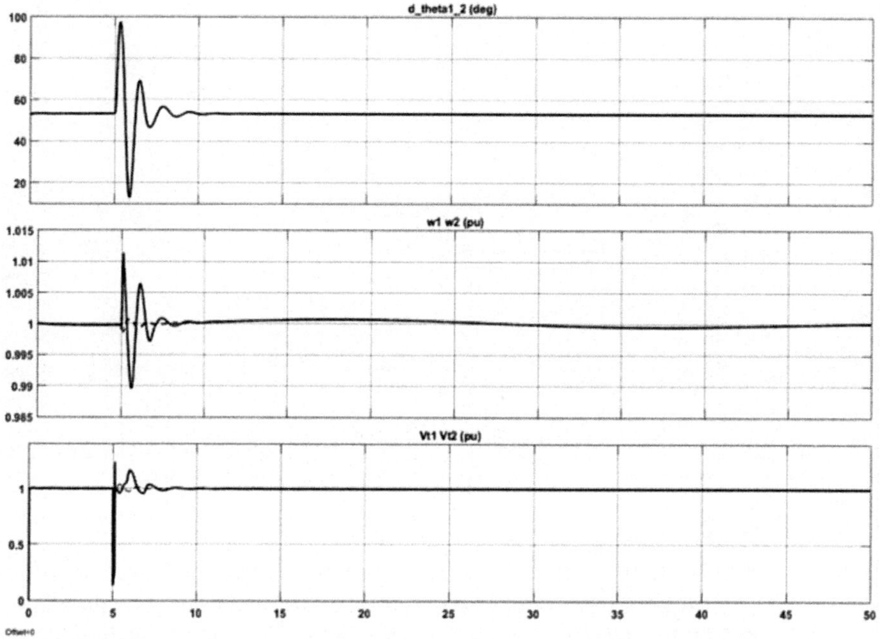

Fig. 3.4 Relative change, machine variation and terminal voltages (system with no PV)

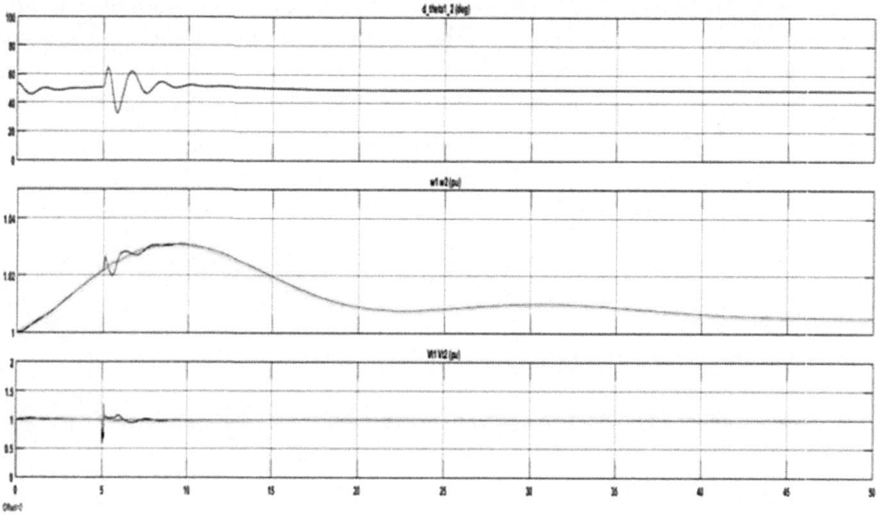

Fig. 3.5 Relative change, machine variation and terminal voltages (with controller)

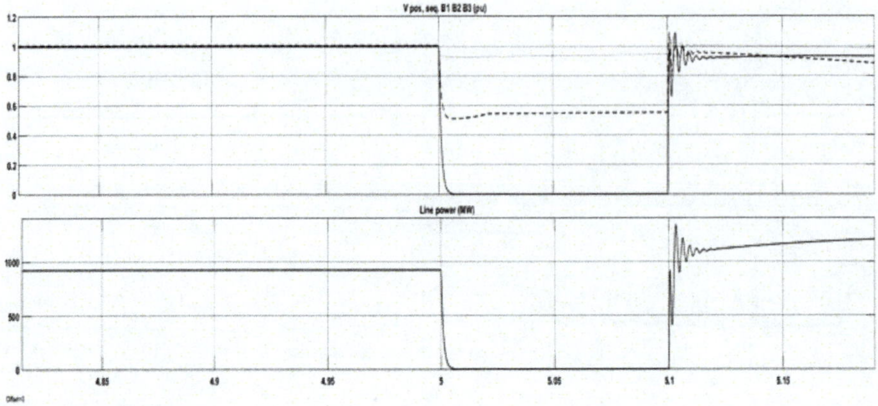

Fig. 3.6 Bus voltages (pu) and line power (MW) (10% PV penetration)

Case-II. Three-Phase Fault

With the controller of 10% solar PV penetration, the statistical measurement of the signal has the maximum speed deviation of 0.00022 (PU) and oscillation duration is 3.23 s; it satisfies the maximum speed deviation at the particular time duration for the stable system. In this case, the particular rotor speed is stable. Bus voltages and line power are shown in Fig. 3.6.

Similar to the above case, system was investigated under 40% PV penetration generation source in the system, and Fig. 3.7 illustrates the inference. The rotor angle peak value further reduced to 92°, and also, electromechanical oscillations are significantly reduced. This signal is a more positive symptom of transient stability improvement. Based on the statistical measurement of the signal, the maximum

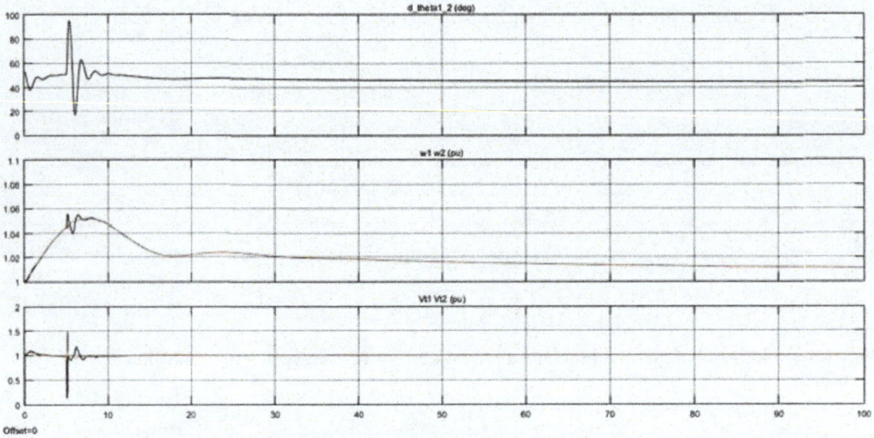

Fig. 3.7 Relative rotor angles, machine speeds and terminal voltages (40% PV penetration)

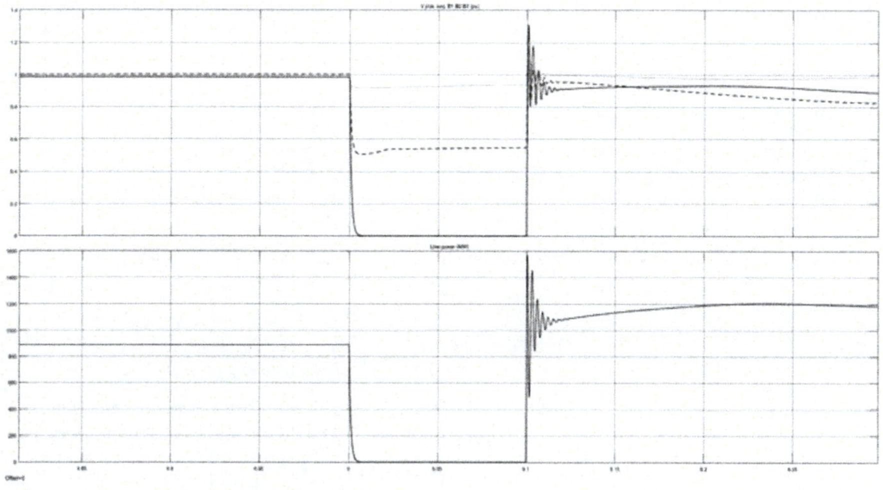

Fig. 3.8 Voltages (pu) and the power (MW) (40% PV penetration)

speed deviation is 0.00022 (PU), oscillation duration is 3.07 s, and it satisfies the maximum speed deviation at the particular time duration for the stable system. In this case, particular rotor speed is stable. Bus voltages and line power are represented in Fig. 3.8.

3.5 Conclusion

Thus, the proposed work shows the impact of PV level during the oscillatory state with and without SVC. The dynamic of the test network is simulated under various faults like single-phase and three-phase faults, with and without SVC controller under different level of PV penetrations. A solution for improving transient stability margin was proposed in this paper. It increases the CCT, controls the first angle swing below the 120 and maintains the rotor angle deviation up to the stable case ranges.

Appendix

Simulation System Parameters

Two machine power system parameters	Synchronous M/C_1	3ϕ; 1000 MVA; 13.8 kV (rms); 60 Hz; $X_d = 1.305$ pu; $X_d' = 0.296$ pu; Pm = 0.95 pu
	Synchronous M/C_2	3ϕ; 5000 MVA; 13.8 kV (rms); 60 Hz; $X_d = 1.305$ pu; $X_d' = 0.296$ pu; Pm = 0.81 pu
	Transformer data	3ϕ delta-star; 1000 MVA; 13.8 kV/500 kV
	Transmission lines	3ϕ distributed parameters line length 700 km. All values are given in matrix farm per km; $R[0.01755\ \ 0.2758]\ \Omega$; $L[0.8737\ \ 3.220]$ mH; $C[13.33\ \ 8.279]\ \mu$F
	Load	3 phase RLC load; active power = 1000 MW
Controller	SVC	500 kV (rms); 60 Hz; Q_c (var > 0) 200 MVAr; Q_l (var > 0) 200 MVAr; $T_d(s) = 4$ ms; $X_s = 0.03$ pu; $K_p =$ 0 puB/puV; $K_i = 300$ puB/puV
PV Farm	–	Rated output power—100 MW; open circuit voltage (V_{oc}) of 36.42 V/per-panel; short circuit current (I_{sc}) of 8.09 A/per-panel; irradiance = 1000 w/m^2; $T = 25$ °C; inverter power rating is 630 kW. The efficiency of the PV array—13.30% at standard test conditions; number of series modules-24; number of parallel modules 120

References

1. U.S. Environmental Protection Agency, Renewable Portfolio Standards Fact Sheet (2009) [online]. Available: https://www.epa.gov/chp/state-policy/renewable_fs.html
2. Renewable Energy 2013: A Global Status Report. REN21 c/Ounep, Paris CEDEX 09, France (2012)
3. M. Yagami, S. Ishikawa, Y. Ichinohe, K. Misawa, J. Tamura, Transient stability assessment of synchronous generator in power system with high-penetration photovoltaics (Part 2). J. Mech. Eng. Autom. **5**, 401–406 (2015)
4. N. Srisaen, A. Sangswang, Effects of PV grid-connected system location on a distribution system, in *Proceedings of IEEE Asia Pacific Conference on Circuits and Systems*, Dec 2006, pp. 852–855

5. S. Achilles, S. Schramm, J. Bebic, Transmission system performance analysis for high penetration. Photovoltaics, NREL (2008) (Rep.no. SR-581-42300)
6. B. Tamimi, C. Canizares, K. Battacharya, System stability impact of large scale and distributed solar photovoltaic generation: the case of Ontario. IEEE Trans. Sustain. Energy **4**(3):680–688 (2013)
7. S. Efftekharnejad, V. Vittak, G.T. Heydt, B. Keel, J. Loehr, Impact of increased penetration of photovoltaic generation on the power system. IEEE Trans. Power Syst. **28**(1):893–901 (2013)
8. M. Yagami, J. Tamura, Impact of high penetration photovoltaic on synchronous generator stability, in *XXth International Conference on Electrical Machines* (2012)
9. L. Wang, D.N. Truong, Stability enhancement of DFIG-based off-shore wind farm fed to a multi machine system using a STATCOM. IEEE Trans. Power Syst. **28**(3), 2882–2889 (2013)
10. M.S. El Moursi, K. Goweily, J.L. Kirtly, M. Abdel-Rahuman, Application of series voltage boosting schemes for enhanced fault ride-through performance of fixed speed wind turbines. IEEE Trans. Power Deliv. **29**(1), 61–71 (2014)
11. S. Farzana Begum, G. Ravi Prakash, Transient stability enhancement of PV farm using ANFIS-PID controller. Int. J. Adv. Technol. Innov. Res. **07**(16), 3096–3103 (2015)
12. N. Ina, S. Yanagawa, T. Kato, Y. Suzuoki, Smoothing of PV system output by tuning MPPT control. Electr. Eng. Jpn. **152**(2), 10–17 (2005)
13. N. Karpagam, D. Devaraj, Fuzzy logic control of static var compensator for power system damping. Int. J. Electr. Electron. Eng. 3–10 (2009)
14. P.H. Reza, J.G. Schavemaker, W.L. Slootweg, W.L. Kling, L. van der Sluis, Impacts of distributed generation penetration levels on power systems transient stability, in *Conference in Eindhoven University of Technology*. IEEE Xplore, 25 Sept 2009
15. S. Raja Mohamed, P. Aruna Jeyanthy, D. Devaraj, Novel control of grid-tied solar PV for enhancing transient stability limit at zero and high penetration levels, in *IEEE International Conference on Intelligent Computing, Instrumentation and Control Technologies*, India (2017)

Chapter 4
Comparative Analysis of CHB and CSD Multilevel Inverter

P. Meeravali Khan, Abdul Khadar Shaik, K. Gopi Krishna, and Aslam Shaik

Abstract The power electronics field has been advancing with new techniques and topologies in broad spread integrating renewable energy sources (RES). The accessibility of resources is decaying ultimately paving the path to RES. In this article, aim is to propose a multilevel inverter using renewable energy as a source, providing AC output power for induction motor applications. Here, we develop two topologies of inverters with cascaded H-bridge (CHB) and cascaded switched diode (CSD) to obtain five- and nine-level staircased output voltage. The configuration is triggered using multi-carrier with a phase disposition SPWM method with less number of switches. The comparative study of both five and nine levels with CHB and CSD configurations with and without filter and total harmonic distortion (THD) parameter using MATLAB/Simulink are done.

Keywords Cascaded H-bridge · Cascaded switched-diode multilevel inverter · Total harmonic distortion

4.1 Introduction

Nowadays renewable energy utilization is a hot debate as it can be economical and effective way of power generation in many industries around the world, which depends on coal, nuclear and other sources for electrical energy that are always a threat to environment. Power conditioning system is to be employed in order to integrate with RES. Thus, MLIs are worn to convert input DC to AC which finds

P. Meeravali Khan (✉)
ANU, Guntur 522510, India

A. K. Shaik
Sree Vidyanikethan Engineering College, Tirupati 517102, India

K. Gopi Krishna
Tirumala Engineering College, Guntur 522601, India

A. Shaik
JNTU Kakinada, Kakinada 533003, India

R. Seyezhai et al. (eds.), *Recent Trends in Renewable Energy Sources and Power Conversion*, Springer Proceedings in Energy,
https://doi.org/10.1007/978-981-16-0669-4_4

wide applications in UPS, conveyors, grid synchronization in micro-grids which are installed in locations where it is difficult to lay overhead, underground long transmission lines, e.g., islands.

There are different kinds of multilevel inverters defined based on their topology like flying capacitors (FC), cascaded H-bridge and switched multilevel inverters [1, 2]. The cascaded H-bridge (CHB) multilevel inverter is to use capacitors and switches and requires less number of components in each level. This topology consists of a series of power conversion cells, and power can be easily scaled. The combination of capacitors and switches pair is called an H-bridge and gives the separate input DC voltage for each H-bridge [3]. It consists of H-bridge cells, and each cell can provide the three different voltages like zero, positive DC and negative DC voltages. One of the advantages of this type of multilevel inverter is that it needs less number of components compared with diode clamped and flying capacitor inverters. These are used for high-power applications like AC motor drives as flying capacitors inverter has capacitors to store energy and have advantages like natural balancing of flying capacitors voltage due to redundancies present in phase leg states, evenly distributed switching stress and removing the necessity of a transformer as capacitors do its duty [4].

Advancement in power electronics has put multilevel conceptual in trends, among them are SPWM, SHE-PWM, SVM, which utilize renewable integration system. CHB MLIs are quite suitable for integration with grid [1]. The multi-carrier PWM strategy is used for switching as it results in lower THD in output voltage [2], but this requires more number of switching devices, and hence, size of the inverter increases, so a cascaded switched-diode multilevel inverter is proposed which results in the reduction of switching devices; the spike removal switch which is added at the source allows reverse load current to flow, and hence, output voltage spikes are removed. Thus, the size of MLI is reduced so as to avoid discontinuities and to provide smooth variation in output voltage using passive filter and can be applied for traction system [5, 6].

4.2 Cascaded MLI and Its Operation

The configuration of cascaded type has multilevel direct voltage (DC) converted to required alternating voltage (AC). It consist of less count of PE devices compared to other MLIs. A number of battery cells formed in a series string chain toward produced AC and final output voltage value are $2n + 1$. Single-phase five-level CHB topology is shown in Fig. 4.1. The circuit consist of eight switches in two series allied in H-bridge arrangement, namely S_1–S_4 and S_5–S_8 with two DC sources. The final output voltage of MLI is $V_o = V_1 + V_2$. The waveforms are of stepped type with reduced harmonics compared to conventional inverter.

A circuit diagram of five-level CHB MLI is shown in Fig. 4.1 with switch combination is depicted in Table 4.1. In parallel, nine CHB MLI are performed by 16 number of switches and nine output levels.

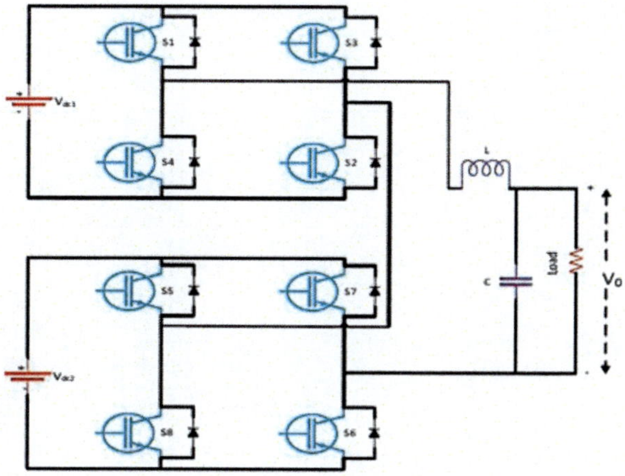

Fig. 4.1 Circuit of CHB five-level topology

Table 4.1 Designing schema for five-level MLI

V_o	S_1	S_2	S_3	S_4	S_5	S_6	S_7	S_8
V_{dc}	1	1	0	0	0	1	0	0
$2V_{dc}$	1	1	0	0	1	1	0	0
0	0	0	0	0	0	0	0	0
$-V_{dc}$	0	0	0	1	0	0	1	1
$-V_{dc}$	0	0	1	1	0	0	1	1

4.3 CSD MLI Operation and Control

Configuration of CSD with five levels is shown in Fig. 4.2. Diodes D_1 and D_2 are connected in parallel with voltage V_1 and V_2 as they keep away from shoot in circuit,

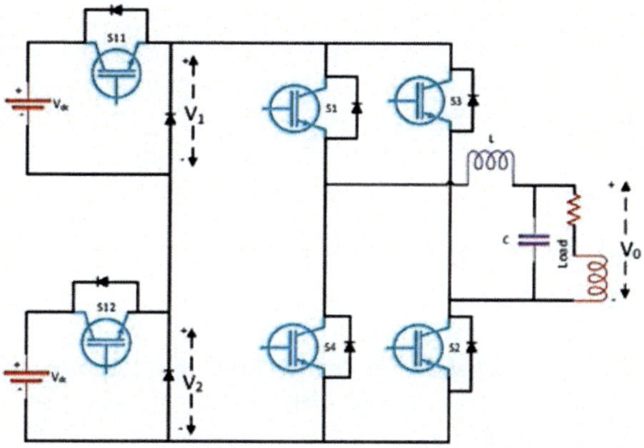

Fig. 4.2 Circuit of CSD five-level inverter

Table 4.2 Switch operation of five-level CSD MLI

V_o	S_{11}	S_{12}	S_1	S_2	S_3	S_4
V_{dc}	1	0	1	1	0	0
$2V_{dc}$	1	1	1	1	0	0
0	0	0	0	0	0	0
$-V_{dc}$	1	0	0	0	1	1
$-V_{dc}$	1	1	0	0	1	1

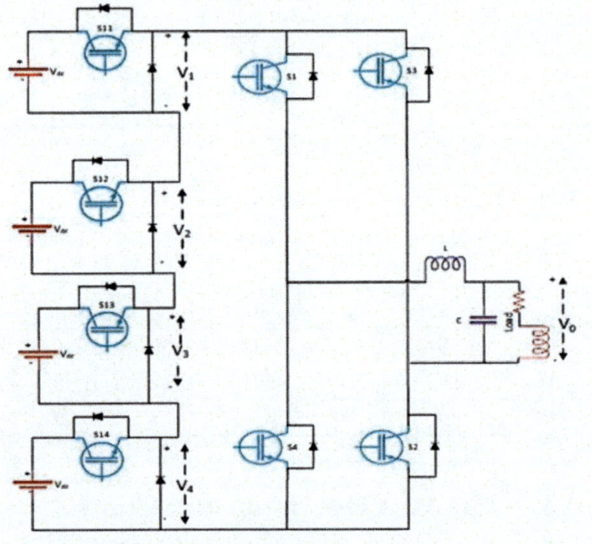

Fig. 4.3 Configuration of nine-level inverter

and the maximum voltage for the first phase is written as

$$V_{01} = V_1 + V_2 + \cdots + V_n \quad (4.1)$$

The timely operation of switches of proposed topology is depicted in Table 4.2. Where as for nine level MLI circuit diagram is shown in Fig. 4.3 and switch operation is Table 4.3.

4.4 Results and Discussion

The voltage levels obtained were $+2V_{dc}$, $+V_{dc}$, 0, $-V_{dc}$, $-2V_{dc}$ and switching sequences are shown in below Table 4.5. LC filter is employed in order to achieve smooth waveform at end with impedance load $Z = 90\ \Omega$, and other parameters are listed in Table 4.4.

Table 4.3 Switch operation of nine-level CSD MLI

V_o	S_{11}	S_{12}	S_{13}	S_{14}	S_1	S_2	S_3	S_4
V_{dc}	1	0	0	0	1	1	0	0
$2V_{dc}$	1	1	0	0	0	0	0	0
$3V_{dc}$	1	1	1	0	0	0	0	0
$4V_{dc}$	1	1	1	1	1	1	0	0
0	0	0	0	0	0	0	0	0
V_{dc}	1	0	0	0	0	0	1	1
$2V_{dc}$	1	1	0	0	0	0	1	1
$3V_{dc}$	1	1	1	0	0	0	1	1
$4V_{dc}$	1	1	1	1	0	0	1	1

Table 4.4 Simulation constrains of five- and nine-level inverter

Design parameters	Five-level MLI	Nine-level MLI
Input voltage	115 V	57.5 V
Output voltage	230 V	230 V
Output current	2.55 A	2.55 A
Load resistor	90 Ω	90 Ω
Filter inductor	35 mH	20 mH
Filter capacitor	5.5 μF	3.5 μF
Operating frequency of inverter	2500 Hz	2500 Hz
Output frequency	50 Hz	50 Hz
Modulation index (m_a)	0.85	0.85

Table 4.5 Switching sequences of five-level MLI

Voltage levels	ON switches
V_{dc}	S_1, S_2, S_5
$2V_{dc}$	S_1, S_2, S_6
0	None
$-V_{dc}$	S_4, S_7, S_8
$-V_{dc}$	S_3, S_4, S_7, S_8

Five-step staircase output voltage is obtainable as shown in Fig. 4.4 with V_i = 115 V and maximum output voltage = 230 V. THD of 28.79% is obtained from analysis as shown in Fig. 4.5 and can be reduced to 2.99% using LC filter as shown in Fig. 4.6.

Nine-level staircase output voltage is available as shown in Fig. 4.7. Input voltage is 57.5 V, and maximum output voltage is 230 V.

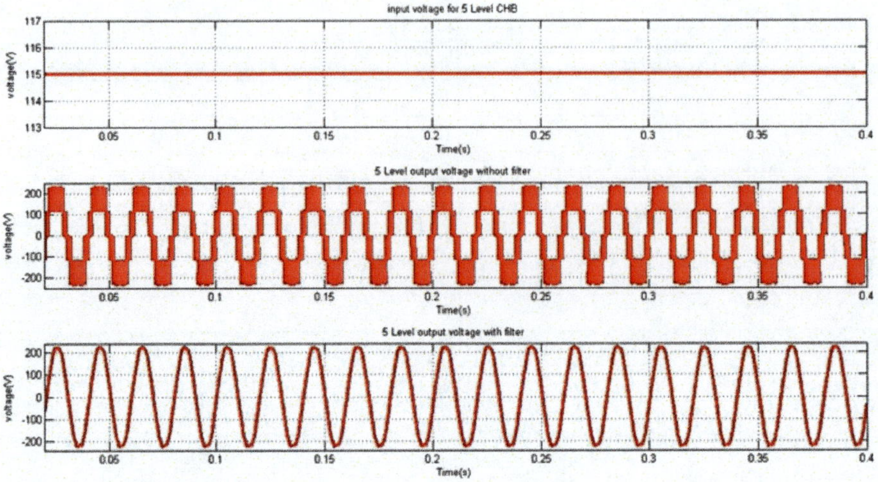

Fig. 4.4 Simulation result of CHB without and with LC filter

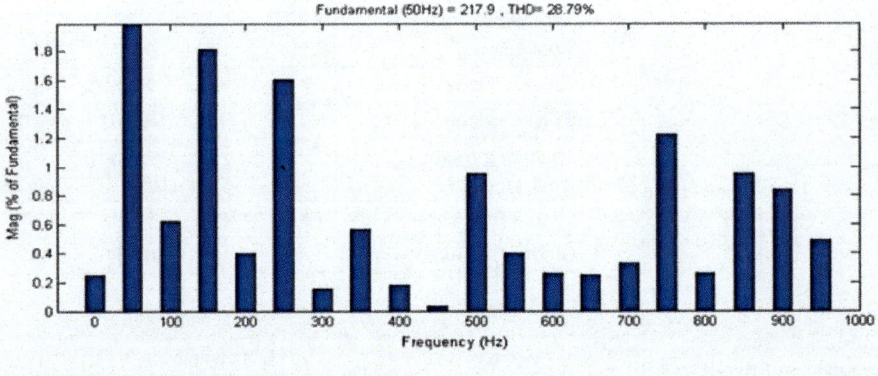

Fig. 4.5 THD of five-level CHB without filter

THD of 14.42% is obtained from above analysis as shown in Fig. 4.8 and can be reduced to 1.66% using LC filter as shown in Fig. 4.9.

The output levels obtained are $-2V_{\mathrm{dc}}$, $-V_{\mathrm{dc}}$, 0, $+V_{\mathrm{dc}}$, $+2V_{\mathrm{dc}}$, to obtain $-V_{\mathrm{dc}}$ the switches S_{11}, S_3, S_4 are switched ON, and to get $-2V_{\mathrm{dc}}$, the switches S_{11}, S_{12}, S_3, S_4 are switched ON. The output voltage for the first stage is shown in Fig. 4.10.

THD of 28.04% is obtained from above analysis as shown in Fig. 4.11 and can be reduced to 1.55% using LC filter as shown in Fig. 4.12.

The output levels obtained are $-4V_{\mathrm{dc}}$, $-3V_{\mathrm{dc}}$, $-2V_{\mathrm{dc}}$, $-V_{\mathrm{dc}}$, 0, $+V_{\mathrm{dc}}$, $+2V_{\mathrm{dc}}$, $+3V_{\mathrm{dc}}$, $+4V_{\mathrm{dc}}$, to obtain $-4V_{\mathrm{dc}}$, the switches S_{11}, S_{12}, S_{14}, S_3, S_4 are in ON condition, for $-3V_{\mathrm{dc}}$, the switches S_{11}, S_{12}, S_{13}, S_3, S_4 are in ON condition, and for $-2V_{\mathrm{dc}}$, the switches S_{11}, S_{12}, S_3, S_4 are turned ON (Fig. 4.13).

THD of 14.66% is obtained from above analysis as shown in Fig. 4.14 and can

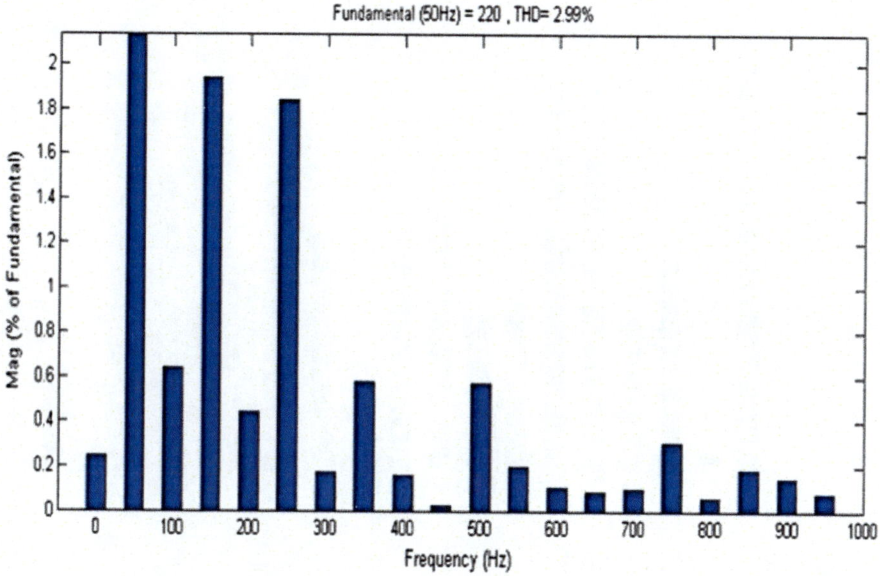

Fig. 4.6 THD of five-level CHB with filter

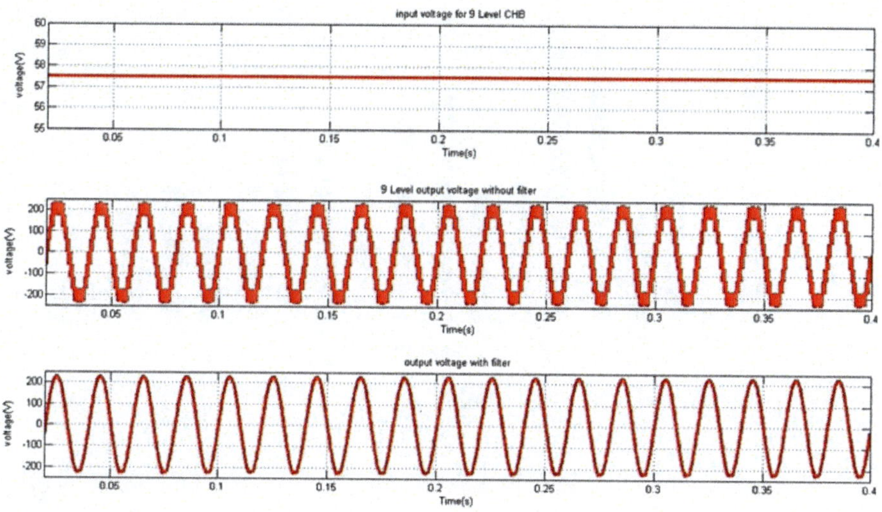

Fig. 4.7 Simulation result of CHB without and with LC filter

be reduced to 0.95% using LC filter as shown in Fig. 4.15. The simulation diagrams of induction motor-fed topology are shown in Fig. 4.16. Comparative analysis for CHB and CSD is shown in Table 4.6 and Fig. 4.17. Clearly illustrates the analysis of

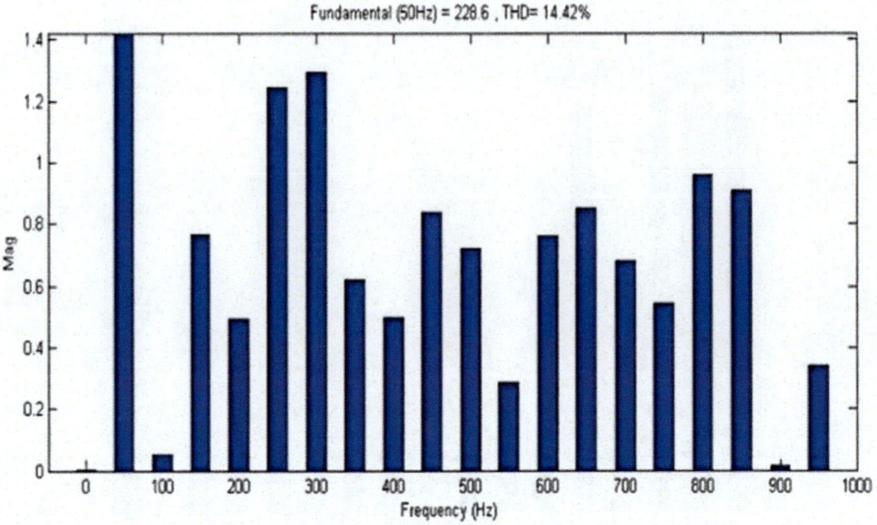

Fig. 4.8 THD of nine-level CHB without LC filter

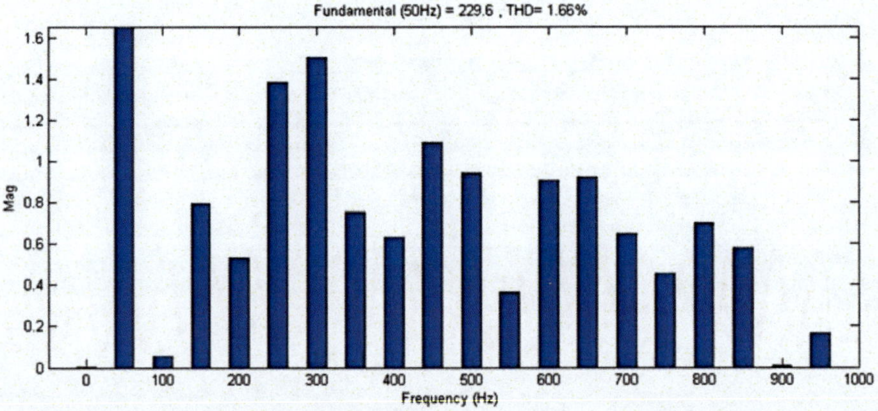

Fig. 4.9 THD of nine-level CHB with LC filter

THD with and without filter and also number of devices required for topology using bar diagram.

The CSD is more efficient when compared with CHB in terms of switches, voltage gain, THD, efficiency and count of switches for the system parameters defined in Table 4.4. The output voltage of 230 V for both configurations is obtained where number of switch count is less for switched diode and harmonic is also very less. The waveforms for five and nine levels with and without filter were drawn. The pictorial representation is shown in Fig. 4.17.

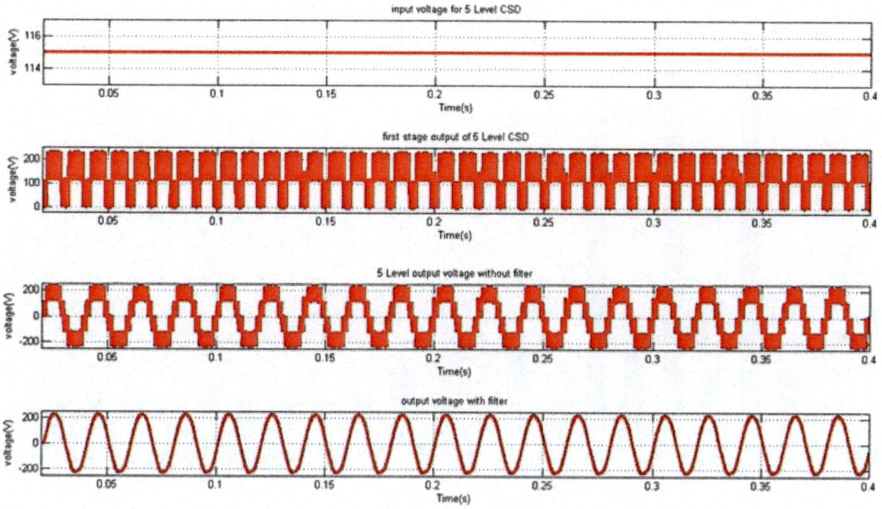

Fig. 4.10 Output waveform of five-level CSD without and with LC filter

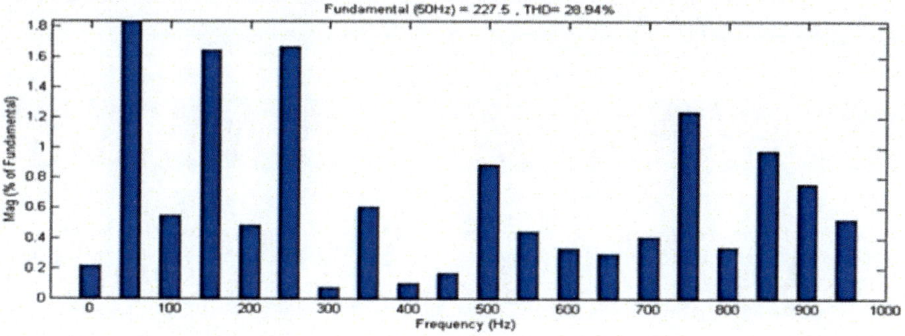

Fig. 4.11 THD of five-level CHB without LC filter

4.5 Conclusion

It has been noted from the simulation result that CSD is more efficient than CHB. The CHB has very high THD, and it is not practical to achieve desired output with such distortion. Comparative study of five and nine levels of both topologies was done with and without filter. From this comparative study, it can be stated that CSD is superior to CSB as it requires less number of switches and gives lower THD in output voltage.

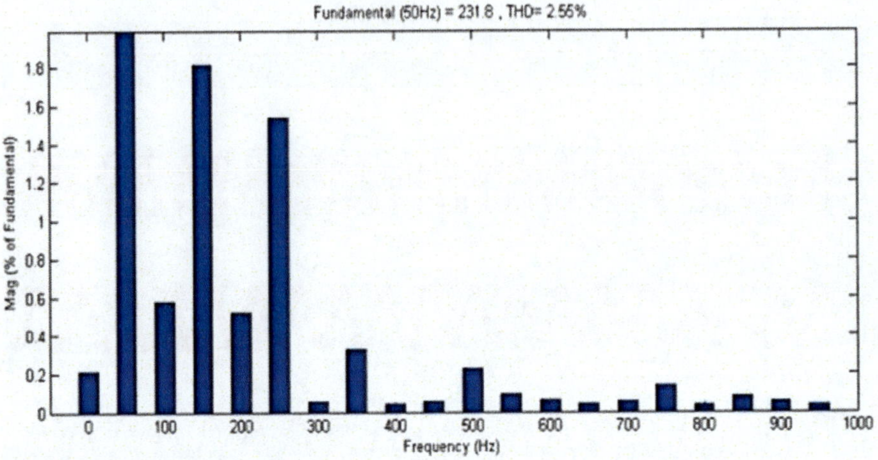

Fig. 4.12 THD of five-level CHB with LC filter

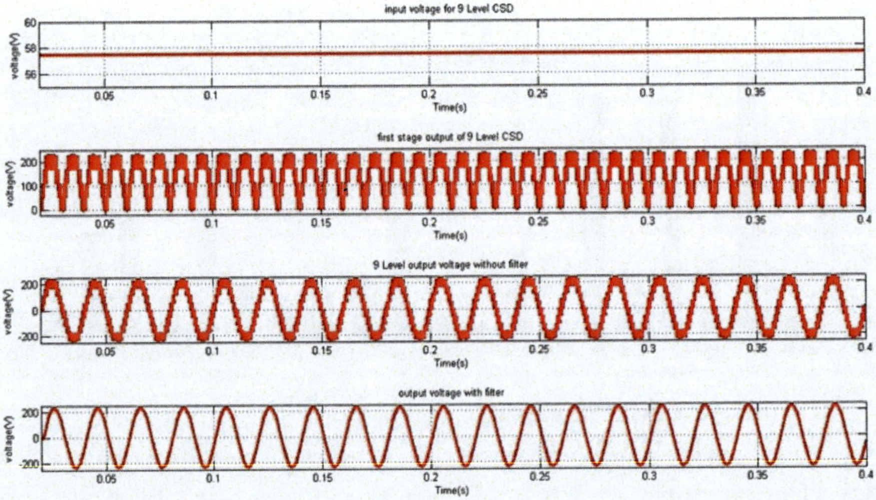

Fig. 4.13 Simulation results of nine-level CSD without and with LC filter

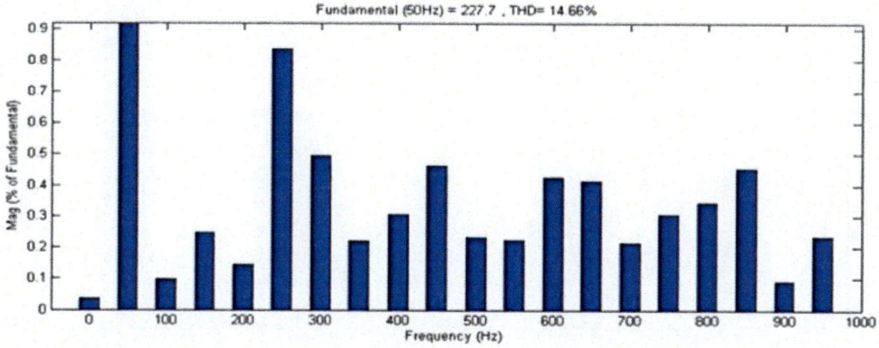

Fig. 4.14 THD of nine-level CHB without LC filter

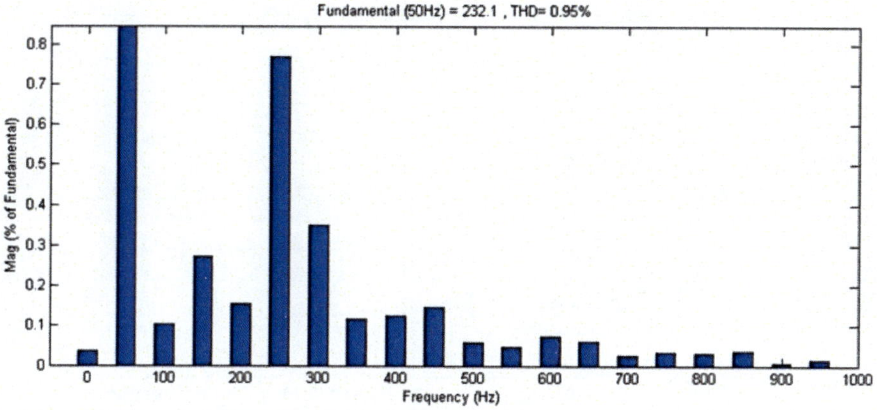

Fig. 4.15 THD of nine-level CHB with LC filter

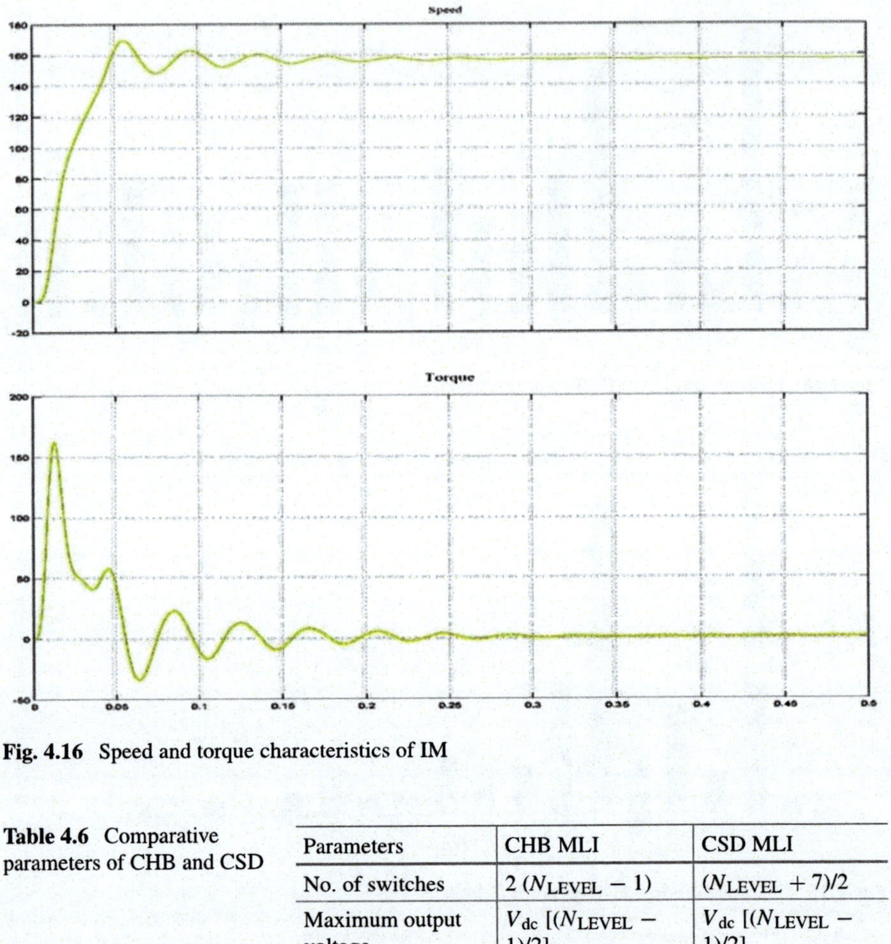

Fig. 4.16 Speed and torque characteristics of IM

Table 4.6 Comparative parameters of CHB and CSD

Parameters	CHB MLI	CSD MLI
No. of switches	$2\,(N_{\mathrm{LEVEL}} - 1)$	$(N_{\mathrm{LEVEL}} + 7)/2$
Maximum output voltage	$V_{\mathrm{dc}}\,[(N_{\mathrm{LEVEL}} - 1)/2]$	$V_{\mathrm{dc}}\,[(N_{\mathrm{LEVEL}} - 1)/2]$

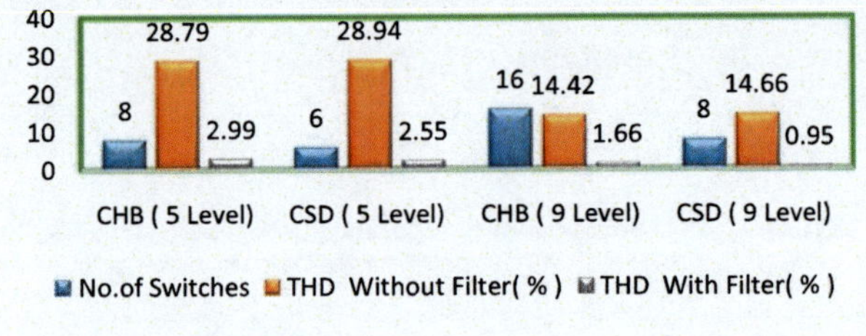

Fig. 4.17 Comparative analysis of CHB and CSD MLI

References

1. K. Gobinath, S. Mahendran, I. Gnanambal, New cascaded H-bridge multilevel inverter with improved efficiency. Int. J. Adv. Res. Electr. Electron. Instrum. Eng. **2**(4) (2013)

2. B.P. McGrath, D.G. Holmes, Multicarrier PWM strategies for multilevel inverters. IEEE Trans. Ind. Electron. **49**(4) (2002)
3. N.A. Azli, Y.C. Choong, Analysis on the performance of a three-phase cascaded H-bridge multilevel inverter, in *Proceedings of the First International Power and Energy Conference PECON*, 28–29 Nov 2006
4. M.S.B. Ranjiana, P.S. Wankhade, N.D. Gondhalekar, A modified cascaded H-bridge multilevel inverter for solar applications, in *Proceedings 2014 International Conference on Green Computing Communication and Electrical Engineering* (2014), pp. 1–7
5. C.I. Odeh, D.B. Nnadi, E.S. Obe, Three-phase five-level multi-level inverter topology, in *Electric Power Components and Systems*, vol. 40 (Taylor & Francis Group, LLC, 2012), pp. 1522–1532
6. J. Swetha, S. Zubeda, Ch.D. Lakshmi Priya, B.M. Reddy, A three phase cascaded H-bridge multi level inverter fed induction motor drive with reduced switches. SSRG Int. J. Electr. Electron. Eng. (ICEEMST'17) (Mar 2017)

Chapter 5
Single Line-To-Ground Fault Protection Scheme for Radial Distribution System

D. Jancy and M. Karthikeyan

Abstract This paper presents a protection scheme against single line-to-ground fault for radial distribution network based on relay agent (RA) to discriminate the fault section. Extraction of informative fault features from zero sequence component is used for discrimination of single line-to-ground fault in the system. To obtain fault features, clusters centers are calculated using fuzzy *c*-means algorithm and using that space relative distance (SRD) is also calculated. The protection scheme is based on the minimum SRD. Finally, coordination of RA is performed to discriminate the faulty section from the system ensuring the better performance without affecting the healthy section.

Keywords Space relative distance · Distribution system · Zero sequence component · Cluster centers · Relay agent

5.1 Introduction

Electrical power distribution system is responsible for supplying power to the consumers in reliable and economical way in that radial networks are used in real time because of its simple structure, design, operation and used only near to the substation and consumers. In such radial networks, single line-to-ground fault is the common type of fault, and it is difficult to identify due to low-level magnitude of fault current. For such system, protection scheme is necessary to detect and isolate the fault section without affecting the entire system before it damages the network. During single line-to-ground fault, the fault current magnitude is low. It makes the detection of such fault difficult.

Over-current relays are used to protect radial networks. Directional over-current relays (DOCRs) are used to protect meshed networks [1].

D. Jancy (✉) · M. Karthikeyan
Department of EEE, Velammal Engineering College, Chennai, Tamil Nadu, India

M. Karthikeyan
e-mail: karthikeyan.m@velammal.edu.in

R. Seyezhai et al. (eds.), *Recent Trends in Renewable Energy Sources and Power Conversion*, Springer Proceedings in Energy,
https://doi.org/10.1007/978-981-16-0669-4_5

In [2], the improved DOCRs protect the radial networks, but this scheme is applicable only for high fault currents. So, it fails to protect against high resistance single line-to-ground fault. Inverse time admittance (ITA) relay is implemented in [3].

The relay operating parameter of ITA relay is the admittance of the line which is not affected by fault current magnitude. But high fault resistance introduces error in admittance calculation making the relay to mal-operate.

Communication networks are used to improve the performance of conventional protection scheme in [1] and [4, 5]. In [2], neural network is used to detect the fault. The input parameters for the neural network are the fault currents. The neural network isolates the faulted zone based on its output. RA isolates the faulted zone in [2]. This RA uses wavelet coefficient of fault current. In all the above methods, a single parameter fault current is used. This will affect the performance of the system under various fault scenarios. An improved method using four parameters to increase the sensitivity and reliability of the protection scheme is implemented in this work. Section 5.2 describes the protection scheme, Sect. 5.3 explains the methodology, and Sect. 5.4 presents the simulation analysis of test system. Section 5.5 presents simulation results and discussions. Section 5.6 presents conclusion of this paper.

5.2 Protection Scheme

According to the protection scheme based on RA, the fault location technique uses below four functions such as

- Fault detection
- Fault features extraction
- Faulted zone identification
- RA coordination.

RA controls the local relay in each line. This RA issues the trip signal to the local relays in the event of fault. This is implemented by sending operate or block signal to the circuit breaker of the line. The RA monitors a minimum of two local relays as shown in Fig. 5.1. This RA can be upstream or downstream based on its location in the system. The RA coordinates with local relays and other RA to detect the faulted line.

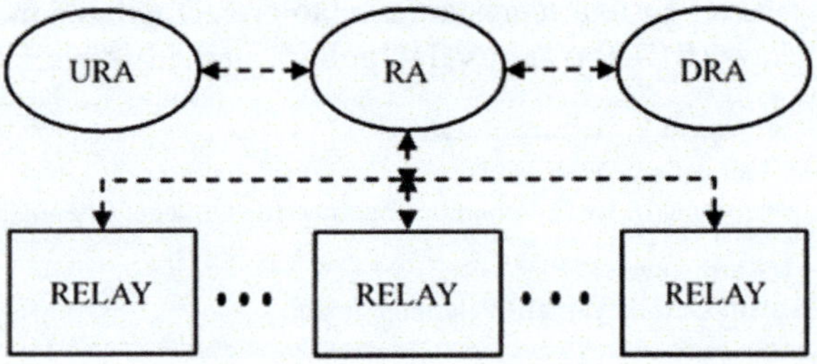

Fig. 5.1 Configuration of RA

According to the framework in Fig. 5.2, the process of protection scheme initiates by collecting the information such as voltage and current.

In the first step, fault is detected using the relay current. If zero sequence current is above thrice the normal current, a fault has occurred as shown in Fig. 5.3. As the result of fault occurrence, increase in the current level causes more damage to the system and equipment which provide inconvenience to the consumers as they are dependent to the single distributor. Therefore, in this step measuring of current level is necessary during fault conditions.

In the second step, extraction of fault features is done once the fault is detected. Having set of information to extract the features with simple steps, these sets of faulted data are normalized by using simple data normalization method.

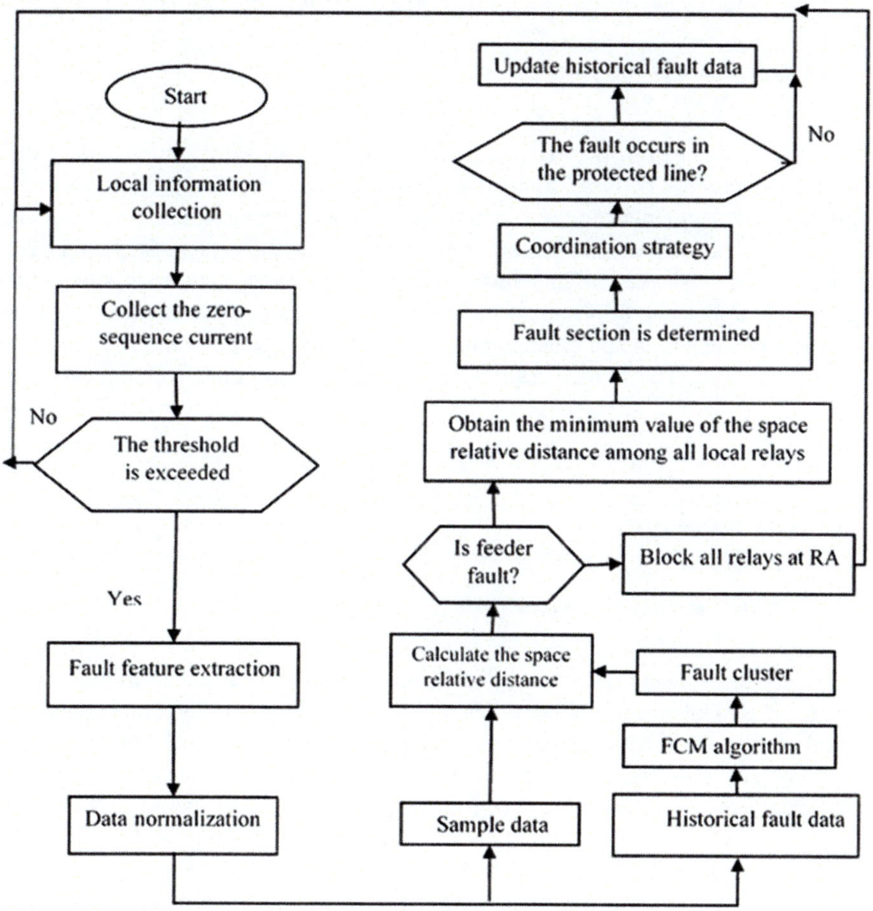

Fig. 5.2 Flow chart of protection scheme

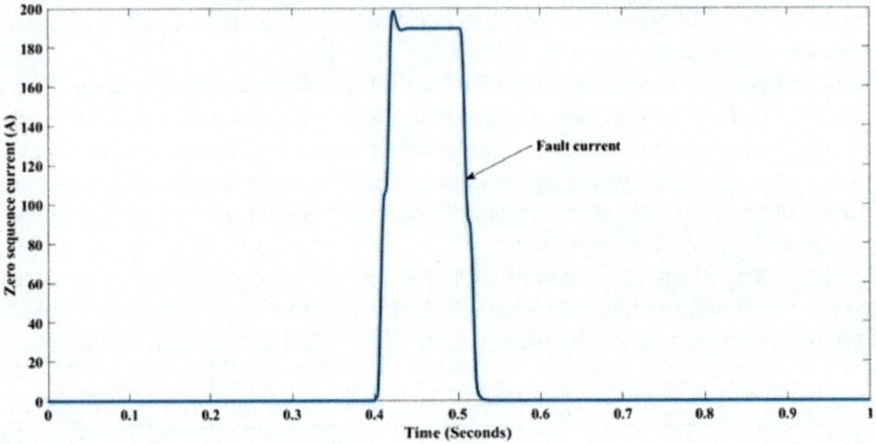

Fig. 5.3 Zero sequence current during fault

For the third step, the obtained fault features are categorized as sample and historic data for the process of fault distinction. Historic data are collected from the past recorded fault accidents and are considered as the fault clusters, and it is processed in fuzzy algorithm for cluster center. The online data are the present fault data which is obtained from the relays. The clusters center shows the space distributions of the fault under various conditions in the network. Distance calculation method calculates SRD between test data and cluster center [6]. SRD tells whether the fault is feeder fault or bus fault. The fault section is identified by minimum SRD among all the lines.

5.3 Methodology

5.3.1 Fault Feature Extraction

The protection scheme detects the fault using features extracted from current and voltage. Based on the number of features and its types, robustness accuracy of fault detection varies [7]. Informative fault features detect the fault accurately. The protection scheme detects fault using zero sequence current. The magnitude of the zero sequence current is higher for ground fault condition compared to the normal operating condition. Thus, protection scheme discriminates the fault from normal condition as shown in Fig. 5.3 for a ground fault.

The protection scheme uses magnitude and phase angle of zero sequence current, phase angle difference between the voltage and current and transient energy as fault features to implement the scheme. When the protection scheme detects the fault, the above features from all the relays of the system are collected. This collected data is named as online data X_s' as follows:

$$X_s' = (I_m, I_{\text{ph}}, \phi_{\text{ph}}, E) \tag{5.1}$$

The stored off-line fault data for each relay are defined as follows:

$$X_f' = \begin{bmatrix} I'_{m11}, & I'_{\text{ph}12}, & \phi'_{\text{ph}13}, & E'_{14} \\ I'_{m21}, & I'_{\text{ph}22}, & \phi'_{\text{ph}23}, & E'_{24} \\ \vdots & \vdots & \vdots & \vdots \\ I'_{mn1}, & I'_{\text{ph}n1}, & \phi'_{\text{ph}n3}, & E'_{n4} \end{bmatrix} \tag{5.2}$$

The protection scheme calculates SRD for the online data X_s with stored fault data X_f for fault distinction. The dataset X' is defined to as follows:

$$X' = \begin{bmatrix} X_s' \\ X_f' \end{bmatrix} = \begin{bmatrix} I'_{m11}, & I'_{\text{ph}12}, & \phi'_{\text{ph}13}, & E'_{14} \\ I'_{m21}, & I'_{\text{ph}22}, & \phi'_{\text{ph}23}, & E'_{24} \\ \vdots & \vdots & \vdots & \vdots \\ I'_{ms1}, & I'_{\text{ph}s1}, & \phi'_{\text{ph}s3}, & E'_{s4} \end{bmatrix} \tag{5.3}$$

The above data have different magnitude range for each feature and have its own effect on the performance of the scheme. *Z*-score method normalizes the matrix in Eq. (5.3). *Z*-score methods eliminate the need of maximum and minimum values for the data. The *z*-score method uses the mean and standard deviation of data. It is described as follows:

$$x_{kj} = \frac{x'_{kj} - \overline{x}'_J}{S(x'_j)}, \quad 1 \le k \le n, \quad 1 \le j \le t \tag{5.4}$$

$$\overline{x}'_j = \frac{1}{n}\sum_{k=1}^{n} x'_{kj} \tag{5.5}$$

$$S(x'_j) = \sqrt{\frac{1}{n}\sum_{k=1}^{n}\left(x'_{kj} - x'_{kj}\right)^2} \tag{5.6}$$

where x_{kj}—normalized value, x_j—mean value of the jth fault feature and $S(x_j)$—standard deviation of the jth fault feature.

The stored fault data X_f are normalized and then converted into the normalized stored fault data X_f. Similar procedure is used for normalizing the online data X_s by using the dataset X' [8].

5.3.2 Fault Distinction

Neural network algorithms [5, 9] are applied for fault detection using online data. In this protection method, space distribution of stored fault data is examined by fuzzy algorithm. The data can be classified as fault or normal current based on the calculated SRD. Fuzzy c-means is one of the best methods for clustering of unlabeled data. Classification of data is based on membership grades. Membership grades are the parameters that indicated the degree of closeness of the data to each cluster. Low membership grades indicate that the data point has lesser degree of closeness to the particular cluster than data in the center of cluster.

FCM algorithm calculates the cluster center for the normalized stored fault dataset X'. The objective function of the FCM is shown as below:

$$J_m(U, P) = \sum_{k=1}^{n} \sum_{i=1}^{c} \mu_{ik}^{m} \|x_k - p_i\|^2, \quad m \in [1, \infty) \tag{5.7}$$

where $U = [\mu_{ik}]c \times n$—membership function matrix and μ_{ik}—membership value of data x_k for cluster i, $p_c = (p_{c1}, p_{c2}, \ldots, p_{ct})$ is cluster center and $c \in [1, n]$—number of clusters, m, greater than 1, is fuzzifier that controls the amount of fuzziness in fuzzy classification, and k is the number of data.

The FCM algorithm can be mainly divided into three steps:

Step 1: For the given c and m, Then membership matrix U is given as follows:

$$\mu_{ik} \in [0, 1] \forall i; \sum_{i=1}^{c} \mu_{ik} = 1, \forall k; 0 \sum_{k=1}^{n} \mu_{ik} \leq n, \forall i \tag{5.8}$$

Step 2: A new cluster center p_i and membership μ_{ik} is calculated as below:

$$p_i = \frac{\sum_{k=1}^{n} (\mu_{ik})^m \cdot x_k}{\sum_{k=1}^{n} (\mu_{ik})^m} \tag{5.9}$$

$$\mu_{ik} = \frac{\left(1/\|x_k - p_i\|^2\right)^{1/(m-1)}}{\sum_k^c \left(1/\|x_k - p_i\|^2\right)^{1/(m-1)}}, \quad \times 1 \leq i \leq c, 1 \leq k \leq n \tag{5.10}$$

Step 3: Based on the new cluster centers p_i and membership μ_{ik}, objective function is calculated. Based on the termination improvement criterion, the FCM algorithm process is stopped. Otherwise, go back to step 2 and repeat the process [6].

5.3.3 Space Relative Distance

Distance calculation method calculates SRD between the nearby data and the remote data. If the data belong to a particular cluster, then the SRD between the data is minimum. Otherwise, SRD is maximum if the data do not belong to the cluster. The various distance calculation methods are as follows:

1. Euclidean distance:

$$d_{kj}^2 = \sum_{j=1}^{t} \left(x_{kj} - p_{kj}\right)^2 \tag{5.11}$$

2. City block metric distance:

$$d_{kj} = \sum_{j=1}^{t} |x_{kj} - p_{kj}| \tag{5.12}$$

3. Chebyshev distance:

$$d_{kj} = \lim_{Z\to\infty}\left(\sum_{j=1}^{t} |x_{kj} - p_{kj}|^z\right)^{1/z} = \max |x_{kj} - p_{kj}| \tag{5.13}$$

5.3.4 Coordination of Relay Agent

Coordination of relay plays an important role in the power system protection. The coordination strategy starts from the measuring unit's RA that collects voltage and current to detect fault along with coordinating with other RAs [7, 10–13].

When a fault occurs, the RA discriminates the faulted line from a healthy line by using the SRD value. The line which is having minimum SRD will be declared as fault line, whereas the healthy lines have higher value of SRD. Figure 5.4 shows the relay coordination strategy implemented by the RA.

5.4 Simulation

Model specification: A radial distribution network with grounding system (10 kV and 50 Hz) as in Fig. 5.5 shows the overhead line L_1, L_2 and L_3. The line parameters include resistance *R*, inductance *L* and capacitance *C* for positive sequence and zero sequence components. The Petersen coil inductance is set as 0.83 H. The values of positive sequence of each transmission line are 0.45 Ω/km, 1.7149 mH/km and

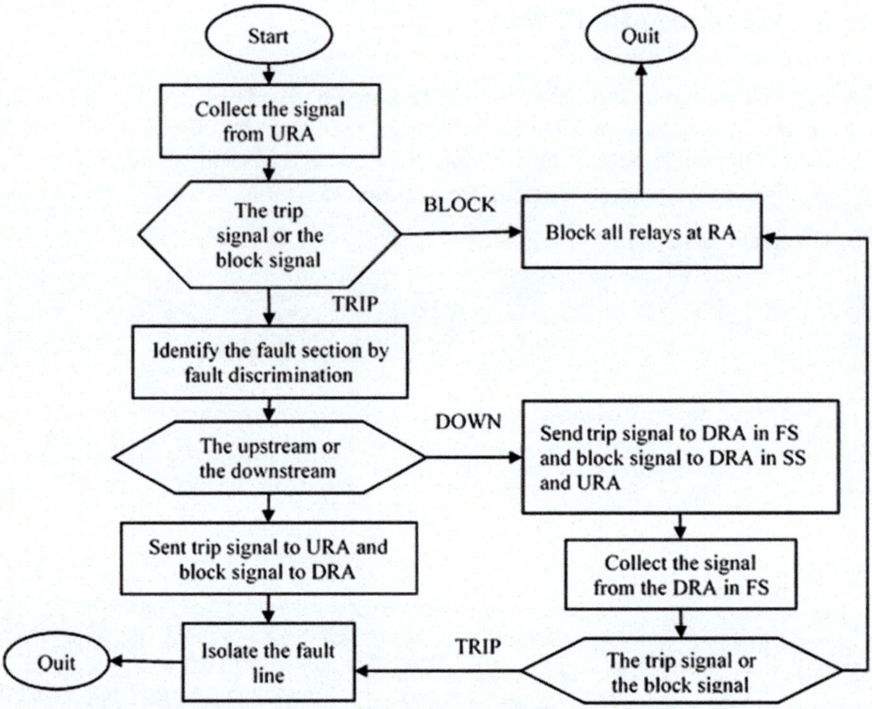

Fig. 5.4 Flow chart of RA coordination scheme

0.9104 μF/km. For zero sequence component of transmission line, the values are 0.7 Ω/km, 3.908 mH/km and 0.3801 μF/km, and along each line, RLC load is used. The test system has four RAs for different buses to control local relays. In the simulation, sampling frequency is 1000 Hz with time step is 1 ms.

Four programed RAs are used in this simulation. RA_1 is used to control the corresponding relays of line 1, line 2, line 3, line 4 and line 6. RA_2 controls relay of line 5. RA_3 controls relay of line 7 and line 8. RA_4 controls relay of line 9 and line 10. Simulation of the radial distribution network is carried out considering the three main parameters: fault resistance, fault inception angle and length of the fault position. Subsystem of each line consists of a measuring unit along with the distributed lines with the length (km) of 10 (L_1), 5 (L_2), 8 (L_3), 6 (L_4), 2 (L_5), 4 (L_6), 4 (L_7), 3 (L_8), 2 (L_9), 5 (L_{10}) and a three-phase fault. For measuring the zero sequence component, sequence analyzer is used. In Fig. 5.6 during fault condition, the current rises twice its normal operation, whereas small changes in voltage can be observed.

The fault conditions are simulated considering different fault conditions. Fault conditions are as follows:

- Fault resistance R_f (Ω): 1, 50, 100
- Fault initial angle δ (°): 0, 18, 36, …
- The percent of the fault position from sending end of feeder L_f (%): 5, 10, 20.

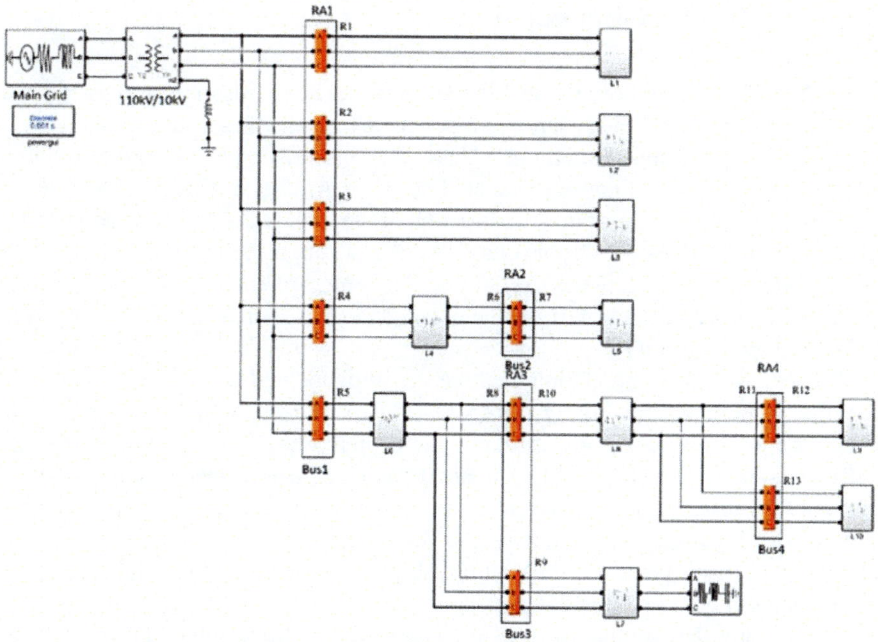

Fig. 5.5 Simulation of test system

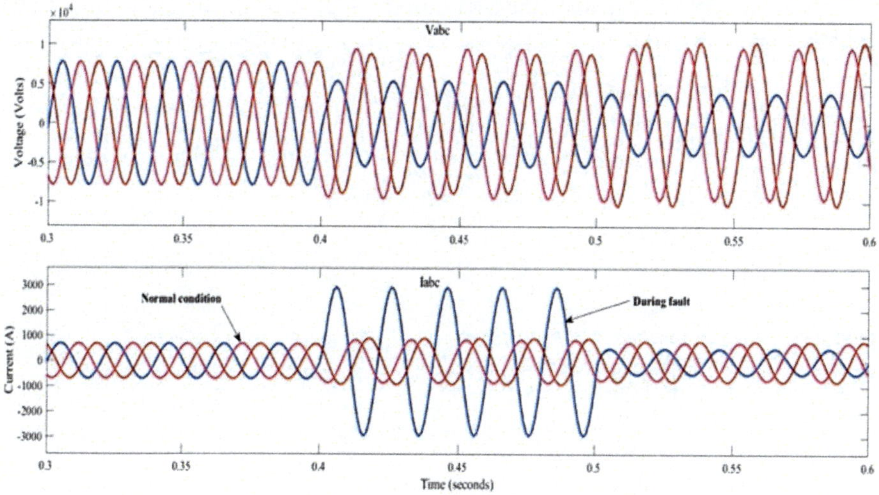

Fig. 5.6 Waveform shows the current contribution during fault condition

5.5 Results and Discussions

In the MATLAB/Simulink, the test system is simulated with time step as 100 ms. The recorded data of zero sequence component are obtained as m-file which is given as input to the programs. Four hundred and fifty samples are generated for each line. All the fault conditions are implemented at phase A of each feeder. The recorded data are utilized in the program for extraction of the fault features by normalizing the data. The extracted features from voltage and current are the stored fault data of every line. FCM algorithm calculates the fault cluster centers of each relay using extracted features as shown in Table 5.1. Each cluster center has four features corresponding to historic data of each relay.

The three distance calculation methods are implemented to find the best distance calculation method using Eqs. (5.10), (5.11), (5.12). The results are shown in Fig. 5.7. On the whole, fault F_1–F_{13} with different fault conditions is corresponding to the line 1–line 10. The SRD using city block method for all relays is shown in Table 5.2 which is the best method for the test system considered. From Table 5.1, the minimum SRD can be seen for each RAs. According to the fourth step of the protection scheme, after the fault takes place, corresponding relays are to be tripped which is marked as underline with bold in Table 5.2.

Considering the simulations, the implemented protection method is effective to identify and isolate the faulty section from healthy section without affecting the network. For example, if a fault F_9 happens in L_9, the first step is that the RA1 initiates the process to calculate the SRD, after the process of calculation from that minimum distance value is obtained for L_6 with respective relay 5 which is identified

Table 5.1 Relays cluster center of stored fault data

Cluster center	Fault features				Relay
	I_m	L_{ph}	ϕ_{ph}	E	
Pf_1	−0.3317	−0.4959	0.5739	−0.2850	1
Pf_2	−0.1316	−0.7848	0.2565	−0.6050	2
Pf_3	−0.1427	−0.8151	−0.3133	−0.6690	3
Pf_4	−0.1953	−0.6754	−0.3554	−0.6431	4
Pf_5	−0.3074	−0.7798	−1.0251	−0.7629	5
Pf_6	0.0196	1.8416	0.9196	1.4583	6
Pf_7	−0.3162	−0.5598	−0.4285	−0.4361	7
Pf_8	0.5142	1.6678	0.6409	1.6936	8
Pf_9	−0.3296	−0.5922	−0.1599	−0.4891	9
Pf_{10}	−0.3317	−0.5982	0.0881	−0.4425	10
Pf_{11}	−0.3246	1.8904	0.7265	1.2658	11
Pf_{12}	−0.3261	−0.3268	−0.3971	−0.3260	12
Pf_{13}	−0.3255	−0.3973	−0.5360	−0.4804	13

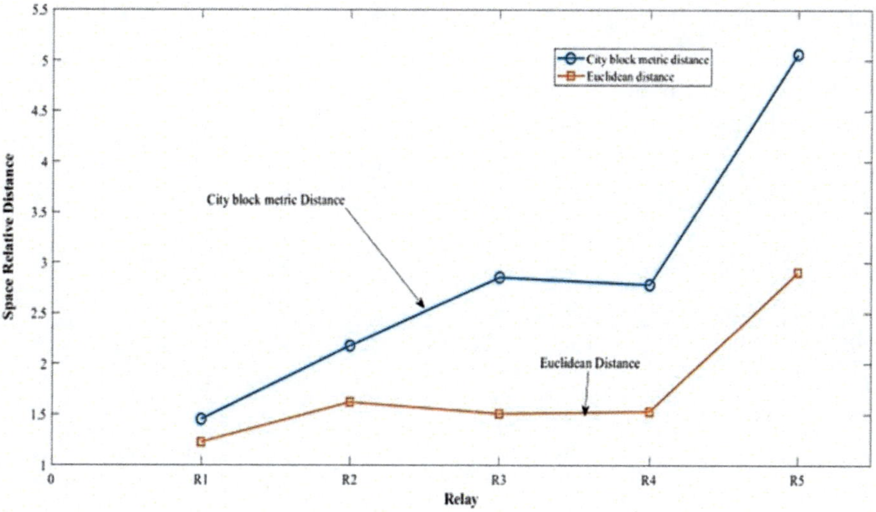

Fig. 5.7 For fault F_1, the SRD obtained using different methods

as faulty section. So that RA_1 sent block signal to RA_2 which makes the corresponding relay under its control and trip signal to RA_3. With that trip signal received, RA_3 performs the same process as RA_1. Hence, the fault line is accurately determined in L_9, and corresponding relay 2 is tripped.

Comparisons are made for the protection method to ensure the better performance even though the fault condition is same for the different line. Table 5.3 shows the coordination of upstream RA and downstream RA for different fault locations. The accurate minimum distance value is obtained which marks as bold one and trip signals are made as underlined one for isolating the faulty feeder from healthy feeder.

5.6 Conclusion

In this paper, a scheme based on RA for single line-to-ground fault protection in the radial distribution networks is implemented for the test system. Detection of fault is identified by zero sequence components, and for the cluster center, fuzzy *c*-means algorithm is used. Different SRD methods are compared. SRD of city block metric distance method is used because of its wide variations between each data than the other methods. Minimum SRD between the online data and cluster center is obtained which indicates the distance of the faulty feeders. Thus, isolating the faulty feeder from the healthy feeder is by performing the coordination procedure. The results of the scheme show the better performance for isolating the specific fault line without affecting the other healthy lines.

Table 5.2 SRD using city block metric method for all relays

(a)

Fault	Fault conditions			RA$_1$				
	R_f (Ω)	δ (°)	L_f (%)	R_1	R_2	R_3	R_4	R_5
F_1	1	0	5	**6.9**	7.0	7.7	7.6	8.6
F_2	50	18	10	1.0	**0.1**	0.8	0.9	1.7
F_3	100	36	20	3.0	1.9	**1.2**	1.4	1.3
F_4	1	0	5	4.2	4.9	4.8	**3.6**	5.8
F_5	50	18	10	9.4	8.6	8.0	7.9	**7.3**
F_6	100	36	20	4.2	4.9	4.8	**3.6**	5.8
F_7	1	0	5	0.8	1.4	1.2	**0.4**	2.0
F_8	50	18	10	7.2	7.8	7.8	8.8	**6.4**
F_9	100	36	20	2.9	2.8	2.3	2.2	**2.2**
F_{10}	1	0	5	2.5	3.1	3.1	4.1	**1.7**
F_{11}	50	18	10	3.6	4.3	4.2	5.2	**3.4**
F_{12}	100	36	20	2.5	2.6	2.1	2.0	**1.7**
F_{13}	1	0	5	3.3	2.8	2.3	2.1	**1.8**

(b)

Fault	Fault conditions			RA$_2$			RA$_3$		RA$_4$		
	Rf_f (Ω)	δ (°)	L_f (%)	R_6	R_7	R_8	R_9	R_{10}	R_{11}	R_{12}	R_{13}
F_1	1	0	5	–	–	–	–	–	–	–	–
F_2	50	18	10	–	–	–	–	–	–	–	–
F_3	100	36	20	–	–	–	–	–	–	–	–
F_4	1	0	5	–	–	–	–	–	–	–	–
F_5	50	18	10	–	–	–	–	–	–	–	–
F_6	100	36	20	**2.3**	4.7	–	–	–	–	–	–
F_7	1	0	5	5.2	**0.8**	–	–	–	–	–	–
F_8	50	18	10	–	–	**2.2**	7.5	7.2	–	–	–
F_9	100	36	20	–	–	6.2	**2.3**	2.5	–	–	–
F_{10}	1	0	5	–	–	4.7	2.7	**2.5**	–	–	–
F_{11}	50	18	10	–	–	–	–	–	**2.9**	4.2	4.2
F_{12}	100	36	20	–	–	–	–	–	5.6	**1.4**	1.5
F_{13}	1	0	5	–	–	–	–	–	6.8	2.1	**1.9**

Table 5.3 SRD for same fault condition in different location

Line	Fault conditions			RA_1				
	R_f (Ω)	δ (°)	L_f (%)	R_1	R_2	R_3	R_4	R_5
L_1	**200**	0	50	**0.50**	0.63	1.28	1.11	1.89
L_1	**400**	0	50	**0.50**	0.63	1.28	1.11	1.89
L_1	**800**	0	50	**0.51**	0.63	1.28	1.10	1.88
L_2	**200**	0	50	1.07	**0.38**	0.87	1.00	1.55
L_2	**400**	0	50	1.05	**0.40**	0.89	1.02	1.56
L_2	**800**	0	50	1.04	**0.41**	0.90	1.03	1.57

References

1. N. Perera, A.D. Rajapakse, T.E. Buchholzer, Isolation of faults in distribution networks with distributed generators. IEEE Trans. Power Deliv. **23**(4), 2347–2355 (2008)
2. H.H. Zeineldin, H.M. Sharaf, D.K. Ibrahim, E. El-Din Abou El-Zahab, Optimal protection coordination for meshed distribution systems with DG using dual setting directional over-current relays. IEEE Trans. Smart Grid **6**(1), 115–123 (2015)
3. M. Dewadasa, A. Ghosh, G. Ledwich, M. Wishart, Fault isolation in distributed generation connected distribution networks. IET Gener. Transm. Distrib. **5**(10), 1053–1061 (2011)
4. H. Zayandehroodi, A. Mohamed, H. Shareef, M. Farhoodnea, A novel neural network and backtracking based protection coordination scheme for distribution system with distributed generation. Int. J. Electr. Power Energy Syst. **43**, 868–879 (2012)
5. N. Rezaei, M.R. Haghifam, Protection scheme for a distribution system with distributed generation using neural networks. Int. J. Electr. Power Energy Syst. **30**, 235–241 (2008)
6. Y. Wang, G. Wei, H. Yang, H. Chen, Z. Ouyang, Novel protection scheme of single line to ground fault for radial distribution systems with distributed generators. IEEE Trans. Power Deliv. **33**(2), 541–548 (2018)
7. Y. Wang, J. Zhou, Z. Li, Z. Dong, Y. Xu, Discriminant-analysis based single line to ground fault protection using improved PCA in distribution systems. IEEE Trans. Power Deliv. **30**(4), 1974–1982 (2015)
8. E. Ogasawara, L.C. Martinez, D. de Oliveira, G. Zimbra, G.L. Pappa, M. Mattoso, Adaptive normalization: a novel data normalization approach for non-stationary time series, in *International Joint Conference on Neural Networks*, vol. 1, Barcelona, Spain (2010), pp. 1–8
9. Y. Wang, X. Zeng, Z. Dong, Y. Huang, Novel protection scheme of stator single-phase-to-ground fault for power transformers. Int. J. Electr. Power Syst. Res. **53**, 321–328 (2013)
10. Y.N. Wang, J.F. Ye, G.J. Xu, Q.M. Chen, H.Y. Li, X.R. Liu, Novel hierarchical fault diagnosis approach for smart power grid with information fusion of multi-data resources based on fuzzy petrinet, in *IEEE International Conference Fuzzy System*, vol. 1, Beijing, China (2014), pp. 1183–1189
11. H. Wan, K.K. Li, K.P. Wong, An adaptive multi-agent approach to protection relay coordination with distributed generators in industrial power distribution system. IEEE Trans. Ind. Appl. **46**(5), 2118–2124 (2010)
12. P.C. Maiola, J.G. Rolim, A multi-agent system for protection coordination of radial systems in the presence of distributed generation, in *11th International Conference on Development of Power System Protection*, vol. 1, Birmingham, United Kingdom (2012), pp. 1–6
13. M.F. Abdel-Fattah, M. Lehtonen, Transient algorithm based on earth capacitance estimation for earth-fault detection in medium-voltage networks. IET Gener. Transm. Distrib. **6**(2), 161–166 (2012)

Chapter 6
Examination and Analysis of Supply Current Control Strategies for Boost-Flyback PFC Converters

J. Anto Sheeba and R. Seyezhai

Abstract In recent years, LEDs are gaining its popularity in many applications due to their characteristics like long lifetime, small size and high luminous efficiency. Since the LEDs are powered from AC source, they must meet the terms and condition as per the International Electrotechnical Commission (IEC) 61000-3-2 Standards regarding the harmonic content and power factor correction (PFC). One interesting topology is by combining the boost topology and the flyback topology to raise the output voltage while maintaining a fair degree of switch voltage stress. This paper explores the study of AC–DC integrated boost-flyback converter (IBFC) for LED lighting applications with various current control techniques to minimize THD and increase the power factor by forming the input current. The performance parameters like power factor of the supply current, distortion factor, displacement factor and THD of supply current are analyzed, and the results are compared with different current control techniques. The design of a 100 W IBFC operating at 100 kHz is presented, and the results are verified.

Keywords Integrated boost-flyback converter (IBFC) · LED · Power factor correction (PFC) · THD · Power factor · Distortion factor · Displacement factor

6.1 Introduction

LED is a profound energy efficient lighting innovation, and it is capable of transforming the future lightning technique. Due to extensive use of LED lighting, the potential effect on energy savings has improved a lot. The quick improvement of LED technology results in more goods and increased production quality, which also leads to lower costs. The high performance and directional character of the LEDs

J. Anto Sheeba (✉)
Rajalakshmi Engineering College, Chennai, India
e-mail: antosheeba.j@rajalakshmi.edu.in

R. Seyezhai
Sri Sivasubramaniya Nadar College of Engineering, Kalavakkam, Chennai, Tamilnadu, India
e-mail: seyezhair@ssn.edu.in

R. Seyezhai et al. (eds.), *Recent Trends in Renewable Energy Sources and Power Conversion*, Springer Proceedings in Energy,
https://doi.org/10.1007/978-981-16-0669-4_6

makes them suitable for many industrial applications. In street lights, garage lighting, walkway and other exterior lighting, frozen case lighting and mobile lighting LEDs are becoming increasing popular.

Traditional light, like luminescent lamps are simple resistor when operate with conventional AC source. It does not require AC to DC power conversion. LED lamps, however, require electric main supply. They need low DC voltage than AC source. So, AC to DC conversion circuit is required [1]. Therefore, the AC is converted into DC either by two stage or by single stage. In two-stage AC–DC converter, the PFC and DC–DC converter are the two stages which are separated by a capacitor bank; whereas, in single stage, PFC and DC–DC converter will operate together by sharing a single switch [2]. Active and passive are the known PFC circuits. An LC filter is placed in passive PFC connecting the AC mains line and the diode rectifier input port. The technique is easy and strong in passive PFC, but the circuit size is larger and heavier, and the power factor cannot be very large. The active PFC controls the quantity of power taken by the load to attain the power factor that is as close as possible to unity. Active components are used in conjunction with reactive element to boost the current shape of the line and to obtain regulated output voltage [3].

In the infrastructure, power supplies connected to AC mains add harmonic currents. Such harmonic currents cause a variety of problems, which distorts the voltage, cause vibration and heating and decreases lines power supply capacity. For effective operation of the utility, "standards" or "recommendations" have forced to use power factor correction (PFC). It is used at the front end to reduce the harmonics. The boost PFC converter is one of the most common active PFC circuits and is a fairly simple and low-cost circuit. Stepping up high voltage can be accomplished by increasing the duty ratio. This causes extreme loss and high-voltage stress in the switching circuit and decreases the performance [4]. The most widely used SMPS circuit is the flyback converter system. It is based on the buck–boost converter and is typically used in applications with low-power consumption. It is important to distinguish the output side from the input side [5].

By integrating the boost converter and flyback converter, another topology called integrated boost-flyback converter (IBFC) shares common switch and the inductor (i.e., a coupled inductor). The coupled inductor of boost topology and flyback converter transformer are combined into one magnetic core to reduce harmonics in the input current and increase the power factor [6]. IBFC gives better efficiency and voltage gain. Very little stress affects the switch and diode.

Nowadays, a greater number of integrated topologies are used to drive LED. A two-stage driver is proposed in [7] to ensure high PF, reduced THD and high efficiency. The magnetic analysis is also performed in order to improve the efficiency. Furthermore, the switching loss and electromagnetic noise can be decreased by making the topology to operate in CCM and by introducing soft switching techniques [8]. By doing so, the weight and size of the converter can be reduced. By introducing an auxiliary winding, THD can be reduced, and the ZVS and ZCS switching will reduce switching loss [9].

This paper introduces the operating theory and theoretical study of IBFC for an application of LED lighting. Various current control techniques are implemented in order to decrease the THD and to get better power factor by shaping the input current. It also compares their performance parameters like total distortion factor (THD), distortion factor (DF), displacement factor and power factor (PF) with various current controller techniques, the necessary waveforms have been presented, and the results are tabulated.

6.2 AC–DC Converter Topology

Modern AC–DC converters are single-stage converter with improved power quality, reduced part count and high performance. One of the most commonly used PFC topologies for LED applications is boost converter and flyback converter.

6.2.1 Boost Converter

Step-up chopper circuit provides high-voltage gain and high efficiency. This can be achieved by boosting the output voltage compared with the lower supply voltage [10]. The circuit diagram is shown in Fig. 6.1.

The relation between output and input voltage is

$$V_{\mathrm{O}} = \frac{V_{\mathrm{I}}}{1 - D} \tag{6.1}$$

Even though the traditional converter attains high efficiency at less voltage, there are several disadvantages with this topology. When the duty ratio is very high, it creates high-voltage stress across the switch, and it causes severe switching loss in power devices. Because the current flowing through the diode is getting reversed, the conduction loss is augmented which corrupts both performance and power level of the step-up converter [11].

Fig. 6.1 Step-up converter

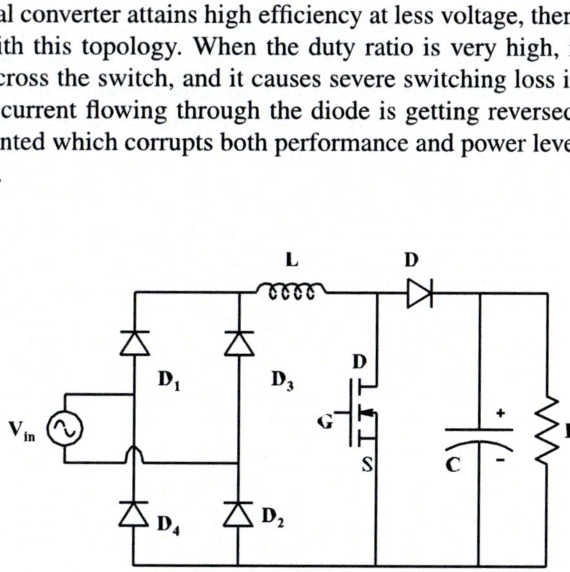

6.2.2 *Flyback Converter*

The most widely used SMPS circuit is the flyback converter. Here the output side and input side are separated. The transformer is used both for isolation and for optimal balancing of input and output voltage with current requirements. This topology is known a "flyback transformer."

Primary and secondary windings in flyback transformer do not conduct simultaneously, and they are like two magnetically coupled inductors. So, naming the flyback transformer as an inductor transformer may be more fitting. The flyback converter topology is shown in Fig. 6.2. Without extreme duty cycle, high-voltage gain could be employed using flyback converter. The voltage gain is determined by the turns ratio of transformer in flyback converter.

The relation between output and input voltage is

$$\frac{V_O}{V_I} = \frac{D}{1-D} \tag{6.2}$$

Still the flyback converter achieves with high-voltage gain, the switching device will undergo high-voltage spike stress. Also, as the voltage gain increases, the efficiency of the converter decreases [12].

6.3 Integrated Boost-Flyback Converter

Switch and diode of this topology are suffered with little stress. High gain and efficiency can be obtained through this converter [13]. In IBFC, the step-up converter performs PFC operation, and the flyback converter will do DC–DC converter operation. Figure 6.3 shows the IBFC. The rectification diode D_6 and filtering capacitor C_2 are present in IBFC. A higher output voltage can be obtained from IBFC by

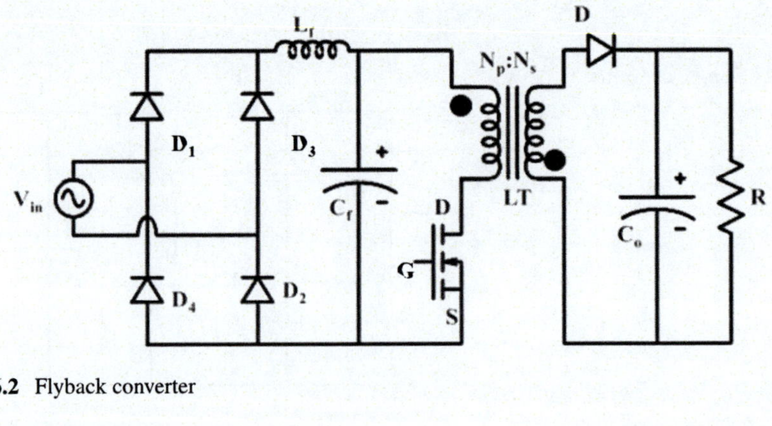

Fig. 6.2 Flyback converter

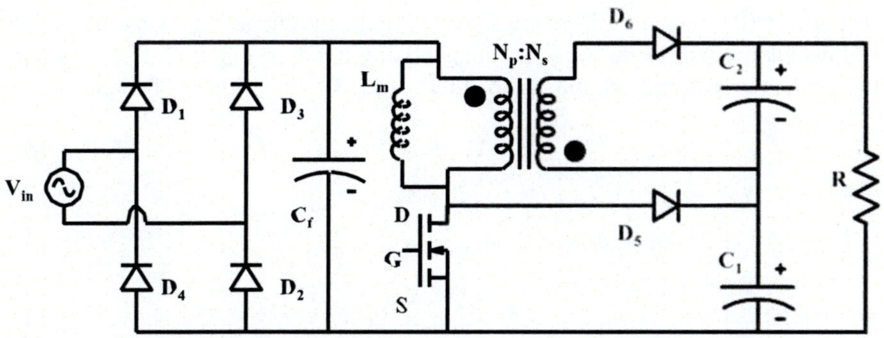

Fig. 6.3 Integrated boost-flyback converter

serially connecting the capacitor (C_1, C_2). The output voltage is higher compared to conventional boost converter without operating the IBFC to extreme high duty cycle. In this, the capacitor C_2 acts as turn OFF snubber capacitor during the turn OFF transient period of switch S in order to suppress the high-voltage spikes. To have a reduced conduction loss, the power switch should be of low rating, and so the overall efficiency can be improved [14].

The coupled inductor and switch S are available in IBFC. The coupled inductor is modeled as an ideal transformer in which both the windings are wound on the same magnetic core. The turns ratio is N_2/N_1, and L_m is the magnetizing inductor. The boost converter output voltage is V_{C_1}, and the flyback converter output voltage is V_{C_2}. The total output voltage is obtained by connecting both the voltages in series connection, thus $V_0 = V_{C_1} + V_{C_2}$. The switch used in the topology is of lower rating, thereby the conduction loss is decreased and the overall output is increased. The operation of IBFC is divided into two intervals.

Model 1: The switch turns ON circuit as shown in Fig. 6.4. During this mode of operation, both diodes D_5 and D_6 are reverse biased. During this mode, the voltage

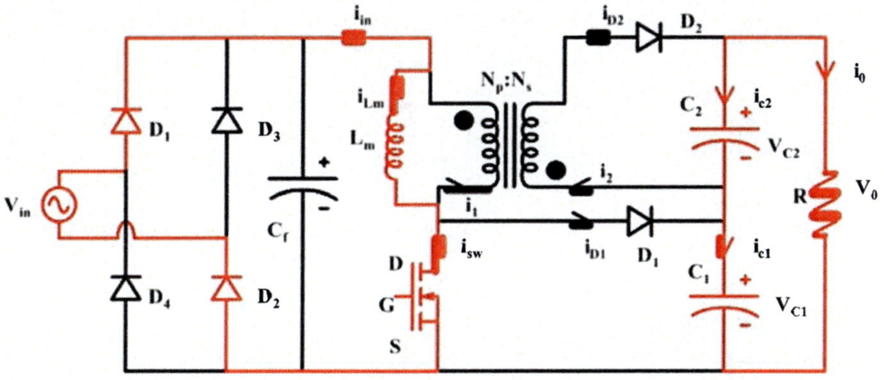

Fig. 6.4 Integrated boost-flyback converter during switch ON

across L_m will be equal to V_i, where V_i is the input DC voltage. i_{L_m} increases linearly, and the energy will be stored in the magnetizing inductor L_m. Both the capacitors, C_1 and C_2, are discharged through load resistance R and providing the load current.

$$V_{L_m} = V_i \tag{6.3}$$

$$I_{C_1(on)} = I_{C_2(on)} = -\frac{V_o}{R} = -I_o \tag{6.4}$$

$$V_o = V_{C_1} + V_{C_2} \tag{6.5}$$

Mode 2: The switch turns OFF circuit as shown in Fig. 6.5. During this mode of operation, both the diodes D_5 and D_6 are forward biased. i_{L_m} reduces linearly, and the energy stored in L_m is transferred to the capacitor to charge it.

$$V_{L_m} = V_{C_1} - V_i \tag{6.6}$$

$$V_{C_1} = \frac{V_i}{1-D} \tag{6.7}$$

$$V_{C_2} = \frac{N_S}{N_P}\frac{D}{1-D}V_i \tag{6.8}$$

$$\frac{V_o}{V_i} = \frac{1+\frac{N_S}{N_P}D}{1-D} \tag{6.9}$$

Selection of the critical magnetizing inductance

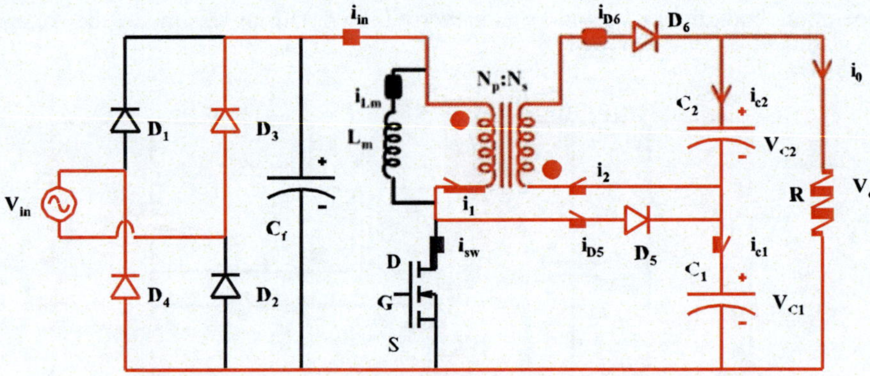

Fig. 6.5 Integrated boost-flyback converter during switch OFF

$$L_{\mathrm{m,Crit}} = \frac{V_o D(1-D)^2 T}{2I_o(n+1)(1+nD)} \tag{6.10}$$

Selection of output capacitor

$$C = \frac{DI_o T}{\Delta v_O} \tag{6.11}$$

6.4 Various Current Mode Controls

In switched mode power supply converter, the main aim was to provide a regulated output voltage for DC loads. Due to the development of many converter topologies and nonlinear loads, it has become a challenge for the converter designers for not only to regulate the output voltage but also to concentrate on wave shape of input supply current when we supply a DC load from an AC main line. For AC–DC power converters, input wave shaping means for a given sinusoidal input voltage the supply current should be sine wave and be in phase with input voltage to increase the power factor. But in open-loop configuration, the power factor obtained will be very less. In order to improve the power factor, the closed-loop techniques can be used. In closed-loop configuration, it consists of two loops. One loop is used to regulate the output voltage, and the other loop is used for shaping the supply current. For power factor correction, the converter can be either operated in continuous conduction mode or discontinuous conduction mode. When we open in continuous conduction mode, the current stress and current ripple will be reduced. This paper makes a comparative study on IBFC with various closed-loop current control techniques like peak current control, average current control and nonlinear current control are simulated, and the results are presented. By using closed-loop current control techniques, the power factor obtained in open loop is improved by wave shaping the supply current and reduces its harmonic distortion.

6.4.1 Peak Control Technique

The control of peak current mode is otherwise called as current programmed control. The closed loop circuit of IBFC using peak current control technique is shown in Fig. 6.6. The switching cycle is triggered by a clock pulse at the set input of the latch, which makes the output of the latch Q as high and so the switch is turned ON.

As the switch is turned ON, the switch current will become same as inductor current, and this current rises with some positive slope depending on the value of the inductance and the voltage of the converter. In due course, both the switch current and control input become same and so the inductor current decreases (refer Fig. 6.7).

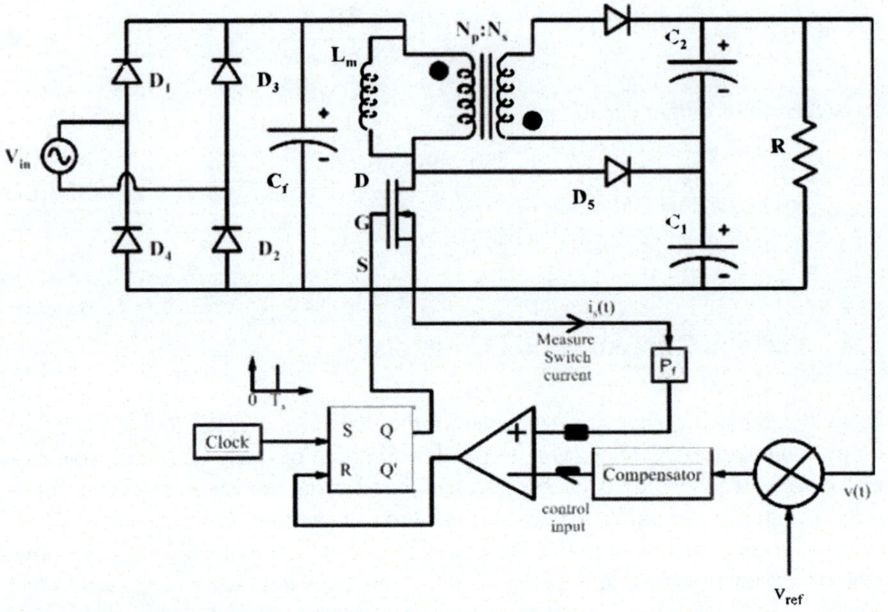

Fig. 6.6 Peak control technique

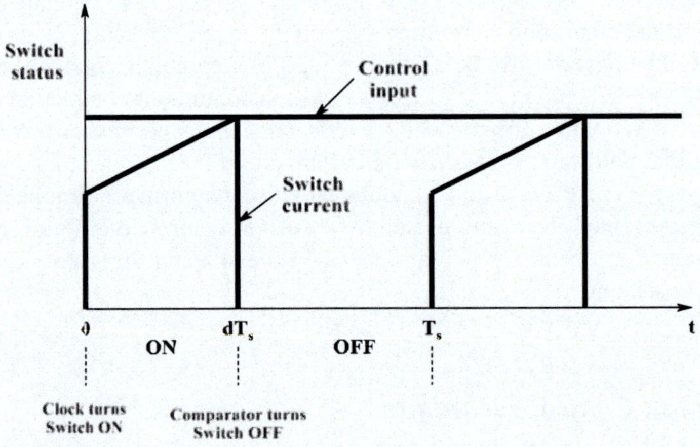

Fig. 6.7 Switch current and control input waveform for the peak control technique

In practice, voltage proportional to switch current and control input is balanced by means of a proportionality constant R_f. If the switching current reaches the control input, the comparator resets the latch and switches the switch OFF for the remaining switching time [15]. The disadvantage of peak current mode control is its weakness to noise in switch current or control input. This noise can cause the latch to be reset too early, and interrupts the controller operation. When the duty cycle exceeds more than 50% the peak current mode control suffers from the issue of sub-harmonic oscillation. This can be resolved by adding of a simulated ramp signal to the sensed switch current.

6.4.2 *Average Control Technique*

The Fig. 6.8 shows the closed loop of average current control technique. It is a well-known strategy for executing input current waveform control in a low harmonic rectifier. This approach can be used in both continuous and discontinuous modes of conduction and can create high-quality current waveform for wide range of input voltages. It is simply a way of controlling two loops. It consists of an internal current loop and outer voltage loop. The output from the error amplifier and outer voltage control loop is then compared with the ramp waveform at the PWM comparator inputs. As this approach is concerned, it is the average inductor current instead of peak current in order to follow the line voltage. When average current mode control is used for power factor applications, the converter works at maximum duty cycle during the zero crossing of line voltage. As a result, the dead angle period found in peak current control is reduced considerably [16]. The key drawback is that the inductor current must be sensed and the current error amplifier needed and the network configuration of the compensator must be taken into account for the different operating points of the converter to concentrate the line period [17].

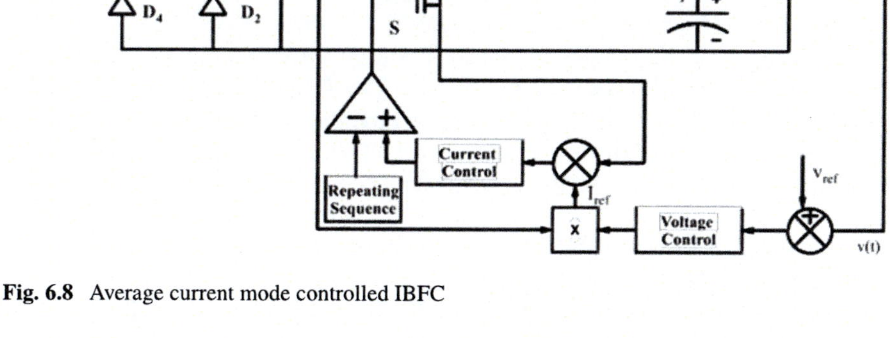

Fig. 6.8 Average current mode controlled IBFC

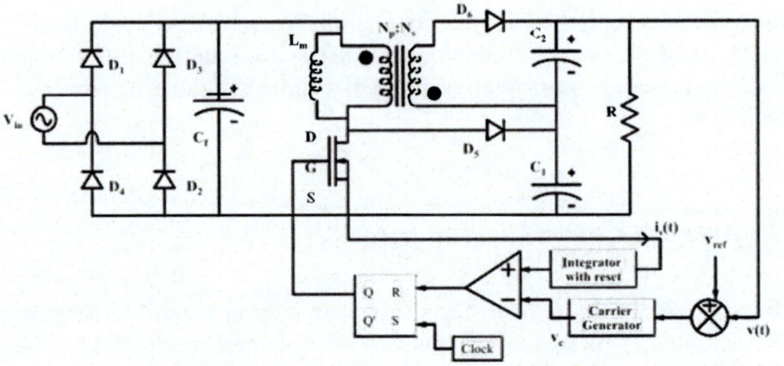

Fig. 6.9 Nonlinear carrier current mode controlled IBFC

6.4.3 Nonlinear Carrier (NLC) Current Control

The nonlinear carrier (NLC) controller does not require the sensing of input current or input voltage. The Fig. 6.9 shows the nonlinear carrier current control technique for IBFC. Current loop error amplifier is also not required like average current control loop. In such a type of controller, the duty ratio is calculated by comparing the switch current with the nonlinear carrier waveform generator. Therefore, the average input current will track the input voltage. For this converter, the duty ratio is obtained by comparing the sensed switch current with the negative ramp carrier waveform for each switching period.

6.5 Software Results

The IBFC is simulated using MATLAB/ Simulink with various control techniques.

The outcome of integrated boost-flyback converter simulation is shown in Fig. 6.10. In Fig. 6.10b, the IBFC THD of supply current under open-loop configuration is 16.66% which is much lower than traditional methods.

The outcome of the Integrated Boost-Flyback Converter simulation with peak current control technique is shown in Fig. 6.11. The THD of this technique is found to be 15.24% which is less than the open-loop configuration.

The result of integrated boost-flyback converter simulation with average control technique is given away in Fig. 6.12. This technique THD is estimated to be 13.99%.

The result of integrated boost-flyback converter simulation with nonlinear carrier control technique is given in Fig. 6.13. This technique THD is estimated to be 12.52%.

The parameters for the simulation are as follows: $V_{\text{in}} = 24$ V, $V_{\text{out}} = 48$ V, $f_{\text{sw}} =$ 100 kHz, duty cycle = 50%, turns ratio 1:2, $L_{\text{m}} = 750\ \mu$H, $C_1 = C_2 = 110\ \mu$F, $R =$ 24 Ω, $P_{\text{out}} = 100$ W. The IBFC performance with various current control strategies is inspected by calculating the performance parameters.

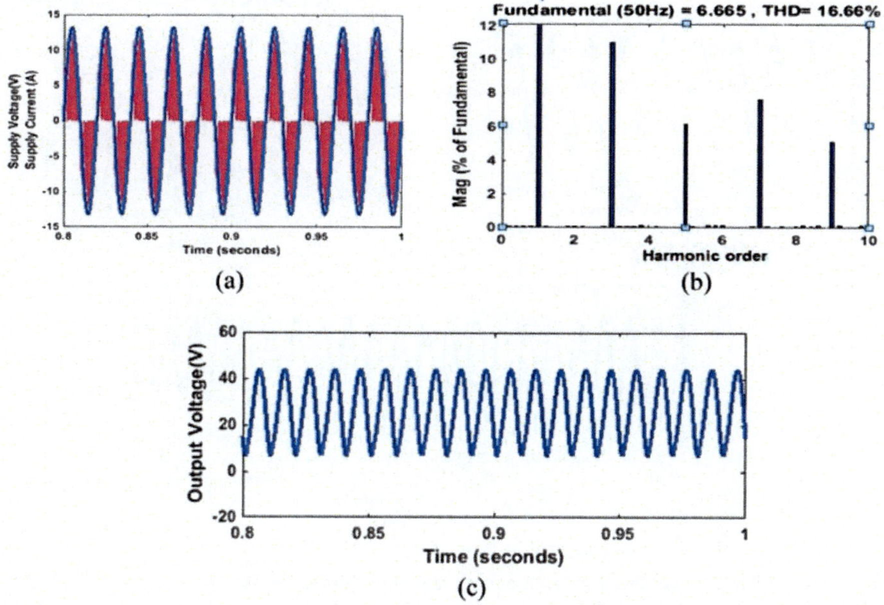

Fig. 6.10 Software result with open-loop configuration **a** input voltage and input current **b** THD of supply current, **c** output voltage

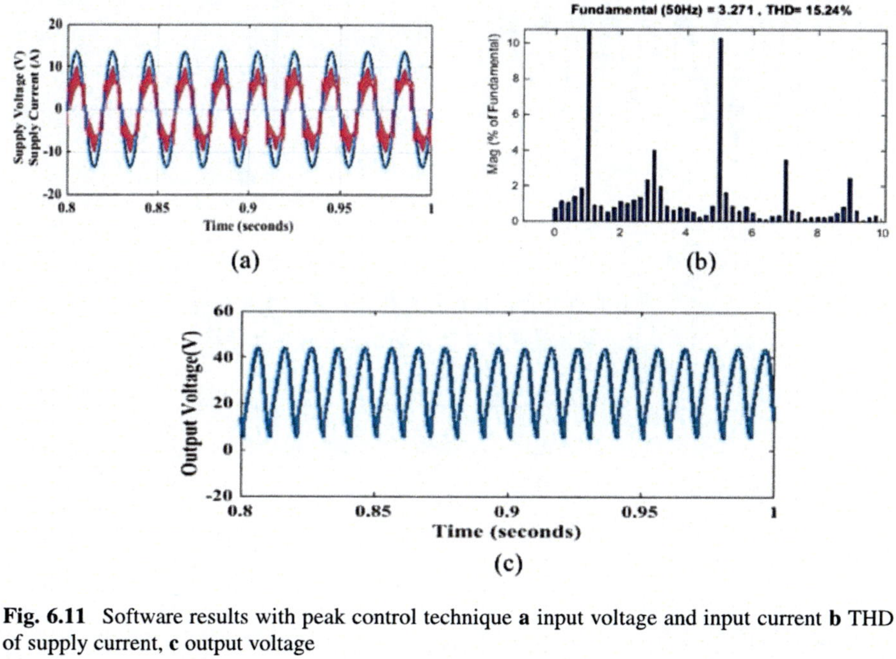

Fig. 6.11 Software results with peak control technique **a** input voltage and input current **b** THD of supply current, **c** output voltage

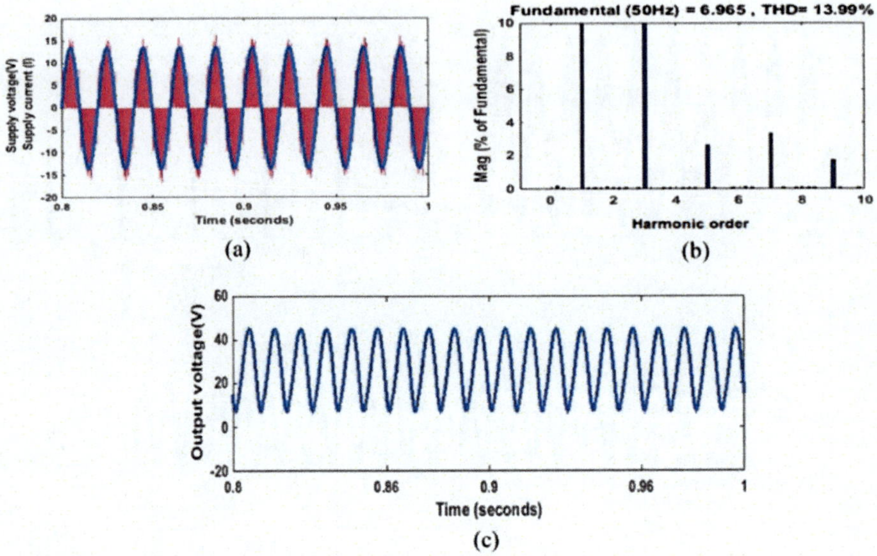

Fig. 6.12 Software results with average control technique **a** input voltage and input current **b** THD of supply current, **c** output voltage

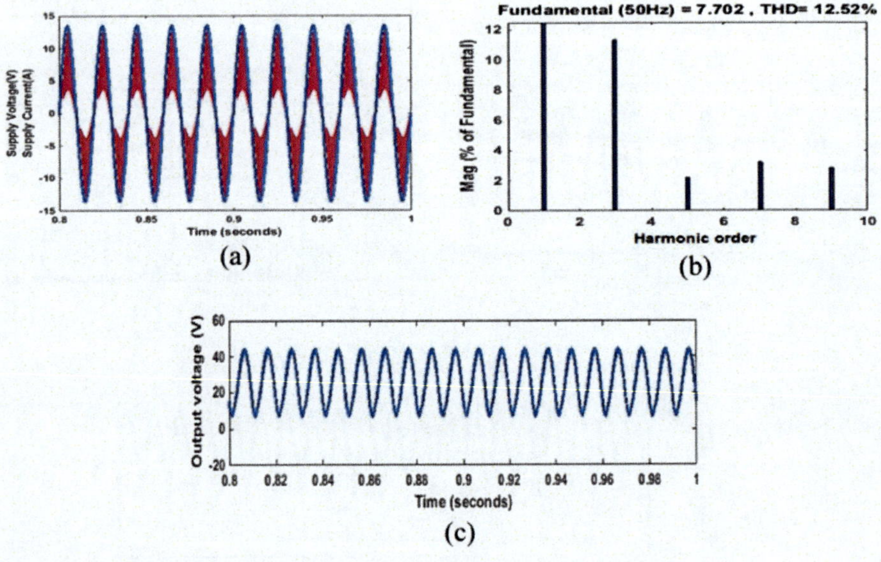

Fig. 6.13 Software results with nonlinear carrier control technique **a** input voltage and input current **b** THD of supply current, **c** output voltage

Table 6.1 Comparison of performance parameters for different current control techniques

Integrated boost-flyback converter	THD (%)	K_d	K_θ	Power factor
Open-loop configuration	16.66	0.941	0.992	0.933
Peak current control	15.24	0.979	0.999	0.978
Average current control	13.99	0.953	0.992	0.946
Nonlinear switch current	12.52	0.960	0.992	0.952

Table 6.1 shows the comparison of THD, PF, distortion factor and displacement factor for different control techniques in integrated boost-flyback converter.

From the above comparison table, THD of nonlinear carrier control gives less value, and the power factor is near unit. Nonlinear closed loop provides better outcome compared to other.

6.6 Conclusion

This paper analyzes the integrated boost-flyback converter with various supply current control techniques. A comparative study of various control methods is performed. From the simulation, it is clear that the IBFC with nonlinear carrier control gives less THD and power factor close to unity. A comparison can also be carried out between Si and SiC devices for better performance.

References

1. J.-H. Kim, D.-Y. Cho, J.-P. Hong, G.-W. Moon, Boost integrated flyback AC-DC converter with valley fill circuit for LED light bulb, in *IEEE Proceeding of the 7th International Power Electronics and Motion Control Conference*, vol. 1 (2012)
2. R.R. Verderver, O.C. Morse, W.R. Alling, Harmonics from compact fluorescent lamps. IEEE Trans. Ind. Appl. **29**, 670–674 (1993)
3. R. Sreemallika, R. Seyezhai, Comparative study of single-stage AC-DC converters for LED lighting with power factor correction. Int. J. Adv. Res. Comput. Commut. Eng. **3**(3) (2014)
4. P. Suresh Kumar, S. Sridhar, T. Ravi Kumar, Design and simulation of boost converter for power factor correction and THD reduction. IJSETR **03**(42), 8462–8466 (2014)
5. J.-J. Lee, J.-M. Kwon, E.-H. Kim, W.-Y. Choi, B.-H. Kwon, Single–stage single–switch PFC flyback converter using a synchronous rectifier. IEEE Trans. Ind. Electron. **055**(3), 1352–1365 (2008)
6. T.J. Liang, K.C. Tseng, Analysis of integrated boost-flyback step-up converter. IEE Proc. Electr. Power Appl. **152**(2) (2005)
7. G.Z. Abdelmessih, J. Marcos Alonso, M.A. Dalla Costa, Y.-J. Chen, W.-T. Tsai, Fully integrated buck and boost converter as a high efficiency, high-power-density off-line LED driver. IEEE Trans. Power Electron. **35**(11) (2020)
8. M. Janbeygloo Esfandani, M. Feizi, R. Beiranvand, CCM operation of a single-stage boost-flyback converter with active-clamp for LED driver applications, in *2020 11th Power Electronics, Drive Systems, and Technologies Conference (PEDSTC)*

9. A. Abasian, H. Farzanehfard, S.A. Hashemi, A single-stage single-switch soft-switching (S6) boost-flyback PFC converter. IEEE Trans. Power Electron. **34**(10) (2019)
10. S. Abdel-Rahman, F. Stückler, K. Siu, PFC boost converter design guide. In-fineon Revision 1.1, 2016-02-22
11. R. Samuel Rajesh Babu, A comparative analysis of integrated boost flyback converter using PID and fuzzy controller. Int. J. Power Electron. Drive Syst. (IJPEDS) **5**(4), 486–501 (2015). ISSN: 2088-8694
12. K.C. Tseng, T.J. Liang, Novel high-efficiency step-up converter. IEE Proc. Electr. Power Appl. **151**(2) (2004)
13. J. Xu, An analytical technique for the analysis of switching DC-DC converters. IEEE Int. Symp. Circ. Syst. **2**, 1212–1215 (1991)
14. R.W. Erickson, D. Maksimovic, *Fundamental of Power Electronics*, 2nd edn. (University of Colorado Boulder, Colorado, 2001)
15. C. Zhou, M. Jovanovic, Design trade-offs in continuous current-mode controlled boost power-factor correction circuits, in *HFPC Conference*
16. L. Rossetto, G.G. Spiazzi, P. Tenti, Control techniques for power factor correction converter, in *Proceedings of the International Conference on Power Electronics and Motion, Control*, Warsaw, Sept 1994, pp. 1310–1318
17. J.H. Lee, J.H. Park, H. Jeon, Series-connected forward–flyback converter for high step-up power conversion. IEEE Trans. Power Electron. **26**(12), 3629–3641 (2011)

Chapter 7
Human Vertebral Spine Segmentation Using Particle Swarm Optimization Algorithm

G. Valarmathi and S. Nirmala Devi

Abstract Upgrade of the spinal segment from computed tomography (CT) and MRI pictures is a pre-preparing step for a scope of picture-guided medications. The objective of this framework is to encourage clinical network to make a quicker pre-screening dependent on the imaging modalities, particularly X-ray and MRI pictures. In this work, MRI of human vertebral spine images has been taken for processing. HE and CLAHE methods are attempted for contrast enhancement of spine images to obtain better visual perception. Contrast difference (CD) is calculated as a performance measure for both HE and CLAHE methods, and the results depicted that CLAHE gives better CD value as against HE. Then denoising is performed using three different filters such as median filter, Wiener filter and bilateral filter. To evaluate the denoising techniques, metrics such as PSNR, SNR and MSE are computed. From the results, the PSNR and SNR values of bilateral filter outperform the other two filters. Diverse edge location strategies are applied for filtered spine images. Performance metrics such as PSNR, SNR and MSE are calculated for those edge detection techniques. Particle swarm optimization (PSO) provides an exquisite method to show the segmentation issue. The dice coefficient is calculated and found to be higher for PSO images which give better segmentation for diagnosis.

Keywords Magnetic resonance imaging (MRI) · Median filter · Wiener filter · Bilateral filter · Canny edge detection · Sobel edge detection · Prewitt edge detection · Robert edge detection · Contrast difference (CD) · Particle swarm optimization (PSO)

G. Valarmathi (✉)
Electronics and Communication Engineering, Sri Sairam Institute of Technology, Chennai, Tamil Nadu, India
e-mail: valarmathi.ece@sairamit.edu.in

S. Nirmala Devi
Electronics and Communication Engineering Department, Anna University, Chennai, Tamil Nadu, India
e-mail: nirmala_devi@annauniv.edu

R. Seyezhai et al. (eds.), *Recent Trends in Renewable Energy Sources and Power Conversion*, Springer Proceedings in Energy,
https://doi.org/10.1007/978-981-16-0669-4_7

7.1 Introduction

Back pain is one of the most well-known medical issues faced by individuals today. It is the second most basic explanation behind a specialist's visit, behind just to the normal virus. Billions of dollars are spent every year on treating back pain, which is additionally an exceptionally regular reason for handicap. Over 90% of individuals will encounter a scene of incapacitating back torment sooner or later in the course of their life. At the point when the problem in the disks has been investigated, the conservative prescriptions like express rest, rubbing power clinical guide or physiotherapy and exercise are suggested. When accurately analyzed, an unnecessary measure of clinical/careful medicines can be kept away from. In clinical imaging, the image processing is used to give good image quality for diagnosis. The human vertebra is a structure consisting of muscles, ligaments, disks and bones which provides support to the body and helps in locomotion. Disks are the space between the segments of spine.

Figure 7.1 pictures the human vertebrae and its divisions. Human vertebra consists of 33 vertebrae with intervertebral disks. Every spinal nerve runs from a particular vertebra in the spinal cord to a particular region of the body. The advances these days empower a fast analysis evaluation contrasted with the manual strategy. The intensity of the image is given by grayscale values of the image. Segmentation of X-ray image based on grayscale values ranging from 0 to 255 is a difficult task.

The image quality is improved by preprocessing technique before segmentation. The essential point in this work is to analyze and evaluate the image enhancement of MRI picture utilizing diverse preprocessing methods. A preprocessing method is basic in image processing. Here two image upgrade techniques are thought about for highlighting the difference and details of the MRI images that are required for clinical analysis. Histogram equalization (HE) is an extremely regular procedure applied

Fig. 7.1 Human spinal cord and its defects

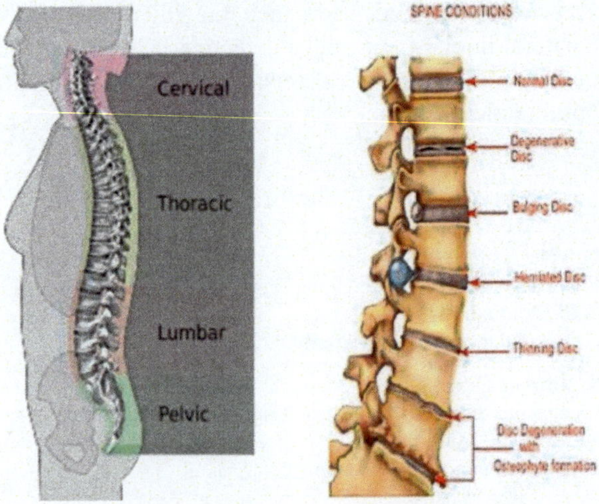

for image upgrade. Contrast limited adaptive histogram equalization (CLAHE) is a well-known image upgrade procedure that changes the gray-level value dependent on certain qualities to highlight the contrast and reduce the noise [1]. Denoising is more significant than any other tasks in image processing, analysis and applications. In this paper, the denoising is done using median filter, Wiener filter and bilateral filter techniques [2]. Different edge detection methods are contrasted with PSO method [3].

Section 7.2 deals with the investigation re-levant to this work, Sect. 7.3–7.7 gives the preprocessing and different methods used in the proposed strategy, Sect. 7.8 talks about the test results and Sect. 7.9 is dealt with the conclusion.

7.2 Related Works

Image enhancement is a procedure to modify the intensity of pixels for improving the segmentation performance. Saenpaen et al. [1] compared three various methods for image enhancement HE, GC and CLAHE method. As discussed in [4], the segmentation for vertebral bone is examined using X-ray images. The two techniques including contrast limited adaptive histogram equalization (CLAHE) and histogram equalization (HE) are discussed. In [5], the X-ray image enhancement, a productive methodology of recognition the dental caries, utilizes the CLAHE method. In [6], the CLAHE technique has a decent histogram esteem, the center worth is high, and the left and right qualities are the less high. Histogram equalization (HE) upgrades the contrast in an image, especially in high-density regions in the histogram. A nonlinear filtering technique—median filter—is utilized on MRI cerebrum images to evacuate salt and pepper noise without disturbing the sharpness of the image. A procedure is where both median and mean filtering [7] are consolidated to expel noise from medical images retaining the structural details. Different filtering methods are compared for spine images out of which bilateral filter performs better than median and Wiener filter [8]. Edge-based techniques permit locale partition through the area of their contours, which have high qualities in an intensity gradient field. A new edge map focusing technique is applied to separate the vertebrae and canal for fusion [9] in which the Canny edge detector performed the extraction of edges with different scales. Canny edge detection method is proved to be good based on average magnitude of image and density of edge length when compared with some medical standard value [2]. Lee applied a 2D changing thresholding technique to recuperate the spine [9]. Boopathy compared the different edge detectors like Roberts, Sobel and Prewitt, Canny and Laplacian of Gaussian with metrics like PSNR, MSE and processing time and suggested Canny detector performs better in spite of higher processing time [10]. Shrivakshan utilized diverse edge detection techniques Roberts, Sobel and Prewitt edge recognition operators, Laplacian-based edge finder and Canny edge detector, and Laplacian-based methods are more easily adaptable to noisy images [11]. Ghamisi utilized PSO and DPSO algorithm for remote sensing data. The processing time and fitness value are better in DPSO method [3].

Hasane Ahammad implemented a nonlinear SVM classification approach based on a hybrid image thresholding method for separation of spinal cord regions. The SCI detection is done using the edge-based nonlinear SVM classifier [12]. A combination of hidden Markov random field and particle swarm optimization method is motivated by the social behavior of the animals, and the performance is evaluated using Kappa index which gives best result compared to K-means and Otsu threshold technique [13].

7.3 Proposed Methodology

In this paper, investigations of two strategies for preprocessing methods for human spine pictures, for example, histogram equalization (HE) and contrast limited adaptive histogram equalization (CLAHE) are applied. The brightness value or tones of the image are depicted in the height of the histogram chart. Each tone has an incentive from 0 to 255, in which 0 is the darkest worth and 255 is the most brightest. It is as yet hard to recognize the right limit in a noisy image. The benefit of the HE is that it works consequently, without the requirement for boundary setting. CLAHE method is a procedure where the full image is separated into rectangular shapes of equivalent sizes, and the adjustment histogram is performed in each squares. The image obtained is high contrast, more the details, and noise is minimized. Contrast difference (CD) is the objective quality measure that takes the advantage of the known characteristics of the human visual system (HVS). According to Michelson formula, [6]

$$\mathrm{CD} = \left| \frac{f_{\max} - f_{\min}}{f_{\max} + f_{\min}} \right| \tag{7.1}$$

$f_{\max}$ and $f_{\min}$ are the maximum and minimum gray-level intensities. $\mathrm{CD} = |C_1 - C_2|$, C_1 and C_2 are the contrast measures of original and enhanced images. Contrast difference (CD) is calculated for the above two methods and compared for spine images in different views in Table 7.2.

Image denoising is performed to remove the noises without disturbing the edges. Here the real patient data are enhanced for gathering important information which helps in clear diagnosis. Image denoising is the traditional way of removing noise in image processing.

This proposed strategy makes use of real patient MRI spine images. Four different patient's MRI spine images are taken in two modalities—sagittal and coronal view for analysis. The images considered are disk degeneration and the normal images. Table 7.1 shows the consequences of applying two enhancement techniques HE and CLAHE strategy for spine images in sagittal and coronal view.

Table 7.2 shows the comparison on contrast difference for HE and CLAHE methods of enhancement techniques applied for MRI spine images. From the experimental results, the contrast difference for the CLAHE method is very less compared to HE where the original information is not lost by enhancement when CLAHE

Table 7.1 Comparison of contrast difference for histogram equalization and CLAHE

	Sagital view		Coronal view	
	Image 1	Image 2	Image 3	Image 4
Original image	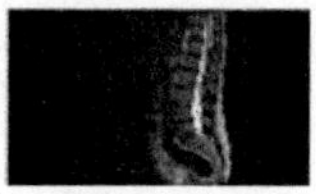	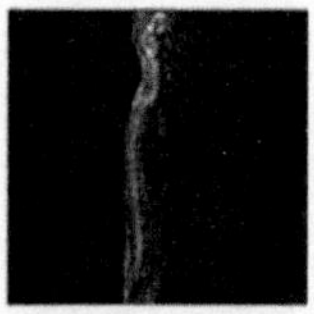	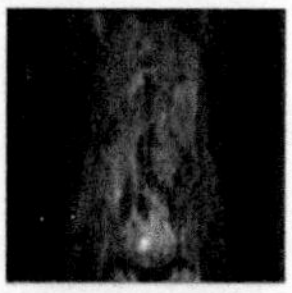	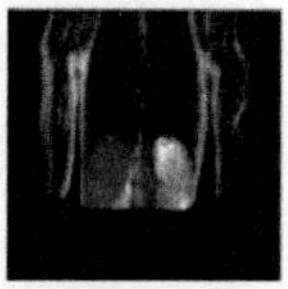
HE image		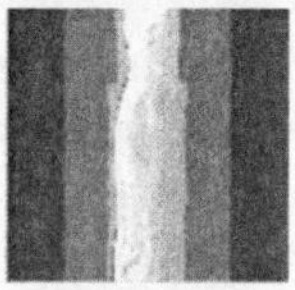	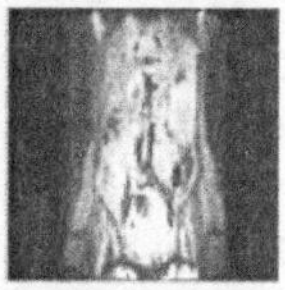	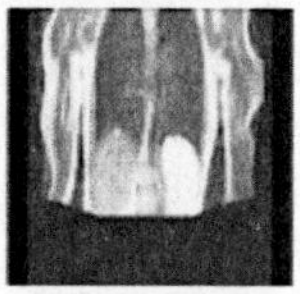
CLAHE image	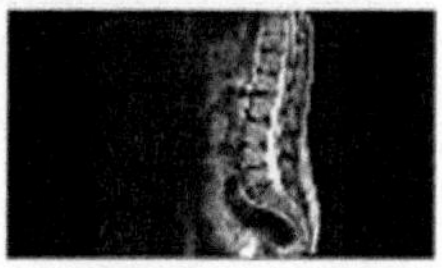	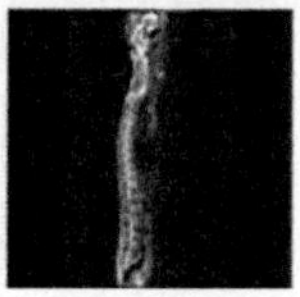	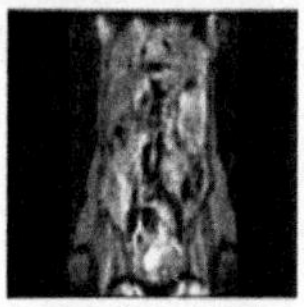	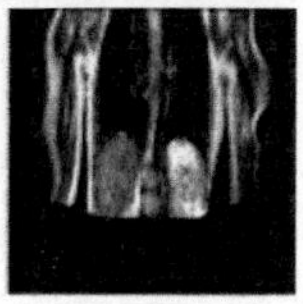

Table 7.2 Comparison of contrast difference for histogram equalization and CLAHE

S. No.	Image	HE	CLAHE
1.	Sagittal view—Image 1	0.3665	0.0003
2.	Sagittal view—Image 2	0.4701	0.1033
3.	Coronal view—Image 3	0.2903	0.0666
4.	Coronal view—Image 4	0.1194	0.0185

method is used. The CLAHE has less contrast difference than HE value, and hence, CLAHE technique has better rebuilding of brightness level which helps in giving more data in diagnosis.

7.4 Different Filtering Techniques

7.4.1 Median Filter

The median filter is a preprocessing technique, often used to remove salt and pepper with edges preserved. Median filter uses a nonlinear operator to replace the pixels in the window with the middle value. All the qualities in the pixel grid from the encompassing neighborhood are estimated from the start, and the pixel under consideration is replaced by the center pixel value.

$$f(x, y) = \text{median}_{uv\varepsilon} K_{xy}\{g(u, v)\} \tag{7.2}$$

where K_{xy} corresponds to the coordinate sets focused at point (*x*, *y*), inside a rectangular sub-image window, and middle value speaks to the middle estimation of the window.

7.4.2 Wiener Filter

The Wiener filter procedure is used to remove noises in images caused by linear movement or unfocussed optics. Wiener filters are created utilizing time domain ideas. They are intended to limit the mean-squared error between their yield and an ideal or required output. The presentation of the Wiener filter might be assessed by listening to signals and noise. The denoising can be done using mean-squared technique, but here least square principle is applied to reduce the mean-squared error (MSE) [2].

$$G(u, v) = F(u, v) \cdot H(u, v) \tag{7.3}$$

where *F* is the Fourier transform of a "perfect" version of a given image, and *H* is the blurring function. The Wiener filter strategy is prepared principally to diminish the noise effect due to blurring effect.

7.4.3 Bilateral Filter

The bilateral filter is a strategy to smooth images while safeguarding edges [8]. The bilateral filter has several qualities as follows:

1. Each pixel is replaced by a normal of its neighbors. In this manner, it makes it simple to acquire intuition about its conduct, to adjust it to application-specific necessities and to actualize it.
2. It relies just upon two parameters that show the size and contrast of the features to save.
3. It very well may be utilized in a non-iterative way. This makes the parameters simple to set since their impact is not aggregated more than a few cycles. Bilateral filter is handled for the most part to lessen the noise effects, thus forming smooth images.

$$\text{BF}[I]_p = \frac{1}{W_p} \sum_{q\varepsilon s} G\sigma_s(\|p - q\|) G\sigma_r\left(I_p - I_q\right) I_q \tag{7.4}$$

Table 7.3 Results for three different filtering methods

	Sagittal view		Coronal view	
	Image 1	Image 2	Image 3	Image 4
Original image	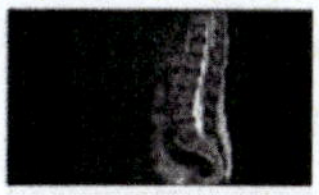	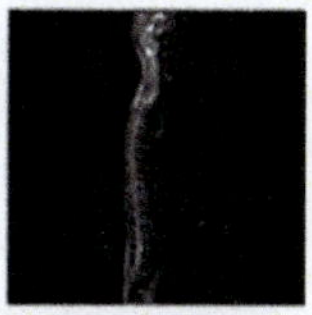	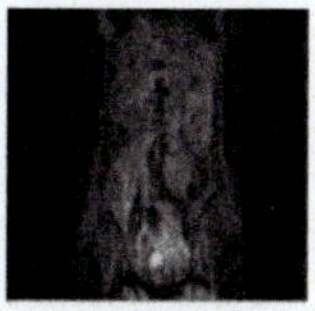	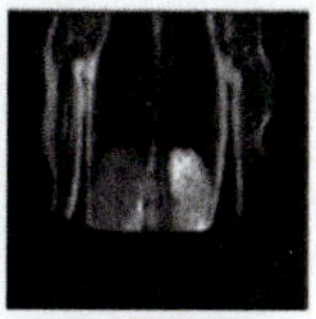
Median-filtered image	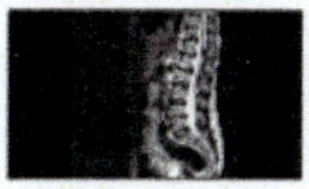	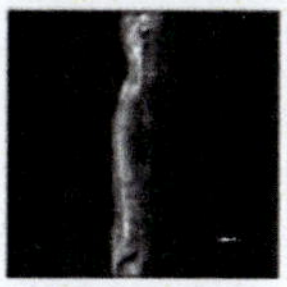	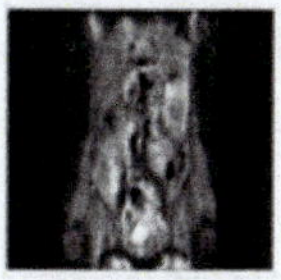	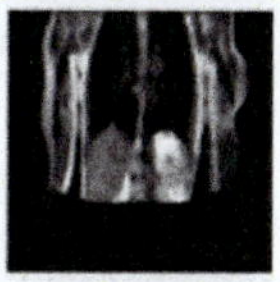
Wiener-filtered image	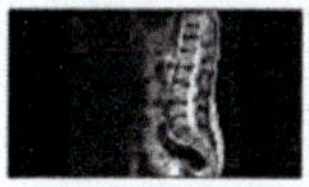	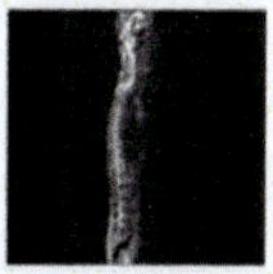	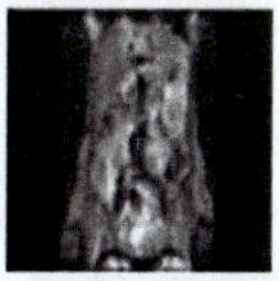	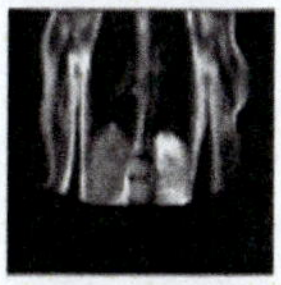
Bilateral-filtered image	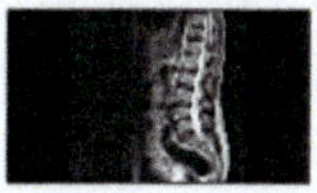	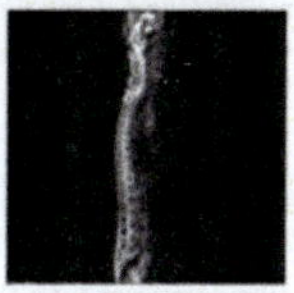	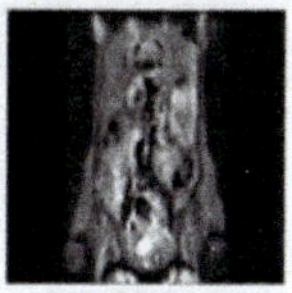	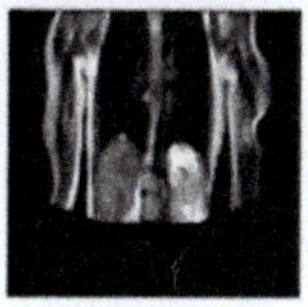

where W_p is normalization factor.

$$W_p = \sum_{q \varepsilon s} G\sigma_s(\|p - q\|)G\sigma_r\left(I_p - I_q\right) \quad (7.5)$$

where σ_s and σ_r are parameters controlling the fall-off of the weights in spatial and intensity domains, which will quantify the measure of filtering for the image *I*, respectively, and denoising of enhanced image is finished utilizing three distinctive filtering techniques. The test results are given in Table 7.3.

7.5 Performance Metrics

The performances of preprocessing procedures were assessed with the presentation measurements, for example, peak signal-to-noise ratio (PSNR), signal-to-noise ratio (SNR) and mean-squared error (MSE) [2].

Peak signal-to-noise ratio is the proportion between the power of a signal and the power of humiliating noise that influences the dependability of its version. The PSNR typically communicated as far as the decibel (dB) scale.

$$\text{PSNR (dB)} = 10\log_{10}\left(\frac{\text{Peak value}^2}{\text{MSE}}\right) \tag{7.6}$$

Peak value is the peak value of the signal.

$$\text{SNR (dB)} = 10\log_{10}\left\{\frac{\sum_{y=1}^{M}\sum_{x=1}^{N}\left[I'(x,y)\right]^2}{\sum_{y=1}^{M}\sum_{x=1}^{N}\left[I(x,y)-I'(x,y)\right]^2}\right\} \tag{7.7}$$

The MSE calculates the normal-squared difference between the estimator and the parameter

$$\text{MSE} = \left(\frac{1}{MN}\right)\sum_{y=1}^{M}\sum_{x=1}^{N}\left[I(x,y)-I'(x,y)\right]^2 \tag{7.8}$$

where $I(x, y)$ and $I'(x, y)$ are the original and processed images.

Some of the performance metrics like PSNR, SNR and MSE are calculated for filtered image and shown in Table 7.4.

A graph is also plotted with these performance metrics values. Figure 7.2 shows the PSNR and SNR values of spine images after filtering process.

From the results, the PSNR and SNR values are high, and MSE values are less for bilateral filter, which preserves edges of the images.

Table 7.4 Comparative measures of PSNR, SNR and MSE

View	Image	Filters	PSNR (dB)	SNR (dB)	MSE
Sagittal view	Image 1	Median	24.5758	13.3978	0.0035
		Wiener	32.4954	21.3174	0.0006
		Bilateral	37.7739	26.596	0.0002
	Image 2	Median	27.6992	12.9786	0.0017
		Wiener	37.6374	22.9168	0.0002
		Bilateral	39.9813	25.2607	0.0001
Coronal view	Image 3	Median	24.6732	15.2667	0.0034
		Wiener	30.3535	20.947	0.0009
		Bilateral	35.4623	26.0557	0.0003
	Image 4	Median	25.4582	15.4303	0.0028
		Wiener	30.278	20.2501	0.0009
		Bilateral	36.5479	26.52	0.0002

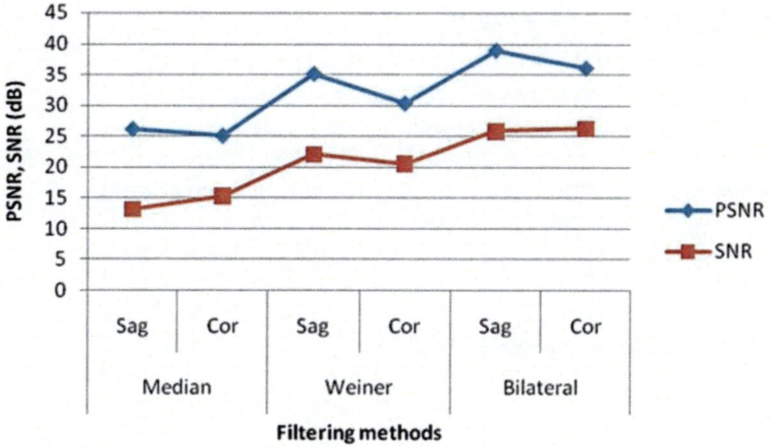

Fig. 7.2 PSNR and SNR of filtered image

7.6 Edge Detection Techniques

Edge detection is mainly used in segmentation and pattern classification of images. There are many edge detectors accessible for pre-handling in images. In any case, Canny, Sobel, Prewitt and Robert's are most applied methods.

7.6.1 Canny Edge Detection

The famous algorithm utilizes four significant procedures to accomplish virtual limit of a digital image. Edges are found by smoothing of images and calculating the derivatives of the pixels. Then the maximum value of the derivatives is calculated, and finally, the thresholding is applied.

The local gradient given by

$$g(x, y) = \left[G_x^2 + G_y^2\right]^{\frac{1}{2}} \tag{7.9}$$

the direction angle given by

$$\alpha(x, y) = \arctan 2\left(G_x, G_y\right) \tag{7.10}$$

are computed at each point, where G_x and G_y are gradients in *X*- and *Y*- axis of the image.

Notwithstanding that, it gives solo edges. Solid edge lines are separated from weak edge points by two-level thresholding method.

7.6.2 Sobel Edge Detection

The Sobel edge detection creates a 2D inclination using Robert cross operator rule. Two 3 × 3 matrices (kernels) are used to find the derivatives at each point and to expand the intensity of the images from light to dim. Every kernel is at an edge of 90°. The edges are focused where the inclination is most noteworthy.

The gradient magnitude is given by

$$G = \sqrt{G_{x^2} + G_{y^2}} \tag{7.11}$$

A rough magnitude is processed utilizing

$$|G| = |G_x| + |G_y| \tag{7.12}$$

The spatial gradient is given by

$$\theta = \arctan\left(\frac{G_x}{G_y}\right) \tag{7.13}$$

7.6.3 Prewitt Edge Detection

The Prewitt edge detection is used to calculate the magnitude and direction of edges in images. The magnitudes in the two axis of image calculate the orientation. For every point in the image, the subsequent inclination estimation is joined to give the gradient:

$$G = \sqrt{G_{x^2} + G_{y^2}} \tag{7.14}$$

The slope's direction is computed by

$$\theta = a \tan 2(G_x + G_y) \tag{7.15}$$

7.6.4 Roberts Edge Detection

The spatial gradient measurement is done using Roberts cross operator. The input operator uses the grayscale image to the result

$$\mathrm{mod}\, G = \left(G_{x_2} + G_{y_2}\right)^{\frac{1}{2}} \tag{7.16}$$

The determined magnitude can be found using:

$$\operatorname{mod} G = \operatorname{mod} G_x + \operatorname{mod} G_y \tag{7.17}$$

Table 7.5 depicts the grayscale images and relating trial resultant edge detected images. Different edge detection methods are applied for four MRI spine images taken in sagittal and coronary view.

The comparative measures of PSNR and MSE for edge detected images in sagittal and coronal view of spine images are estimated and shown in Table 7.6. When the MSE ought to be low, restoration of the images is easier. If MSE values are high, it is guaranteed to discover more edges on the image, and furthermore it is able to identify

Table 7.5 Comparison of various edge detected images

	Sagittal view		Coronal view	
	Image 1	Image 2	Image 3	Image 4
Original image	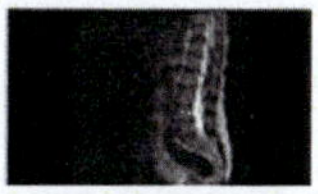	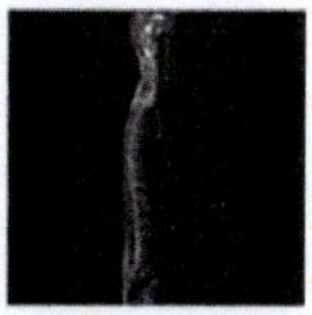	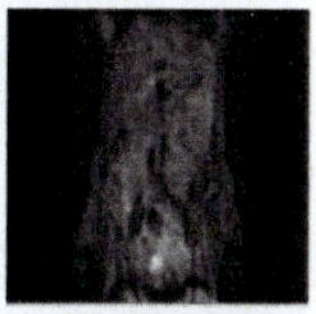	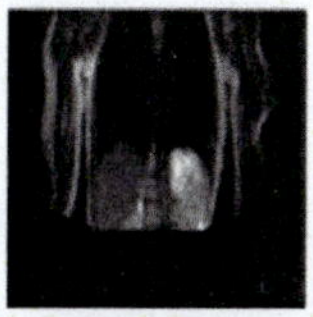
Canny	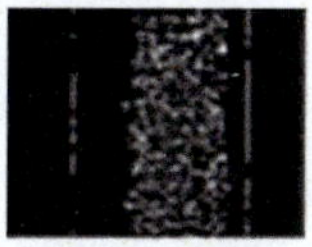	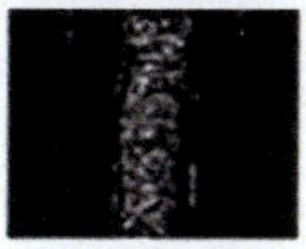	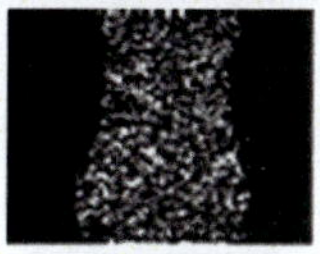	
Sobel		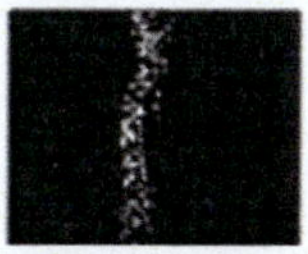	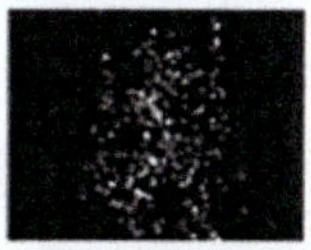	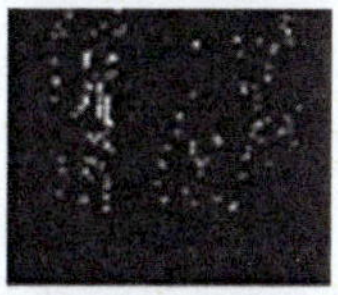
Prewitt		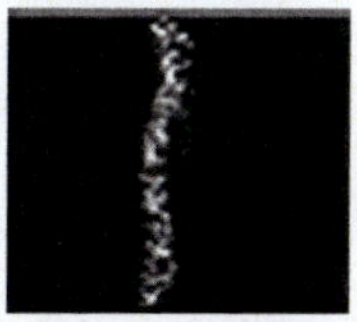	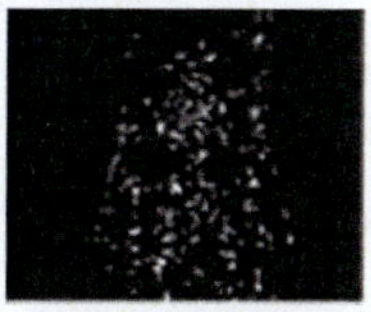	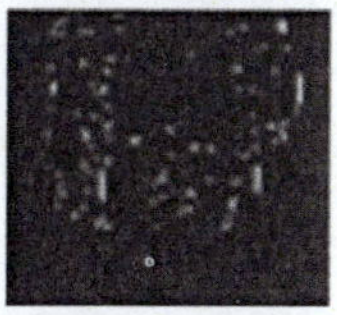
Roberts	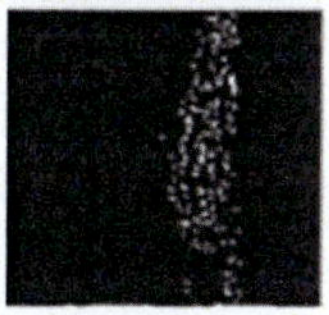	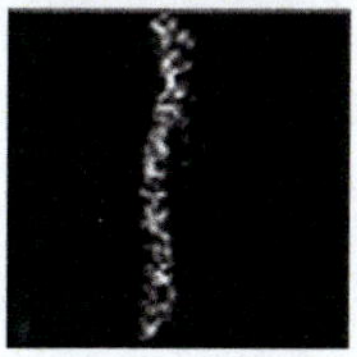	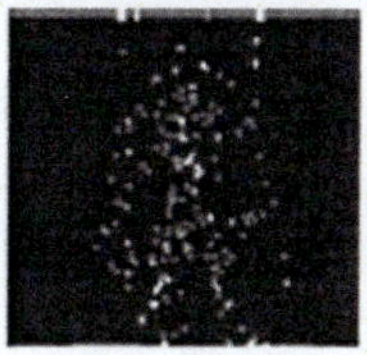	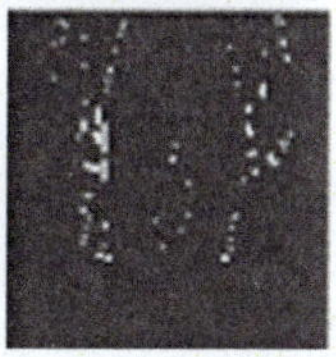

Table 7.6 Comparative measures of PSNR and MSE

View	Image	Edge detection	PSNR (dB)	SNR (dB)	MSE
Sagittal view	Image 1	Canny	10.4463	1.7835	0.0899
		Sobel	11.2344	6.7185	0.0749
		Prewitt	11.2354	6.7451	0.0749
		Roberts	11.3732	6.4959	0.0725
	Image 2	Canny	13.0142	0.6586	0.0500
		Sobel	14.6632	3.2512	0.0342
		Prewitt	14.6629	3.2609	0.0342
		Roberts	14.9877	2.7122	0.0317
Coronal view	Image 3	Canny	8.8865	1.7531	0.1292
		Sobel	9.2144	8.1117	0.1198
		Prewitt	9.2141	8.1570	0.1198
		Roberts	9.3277	8.7521	0.1167
	Image 4	Canny	9.5878	2.8447	0.1100
		Sobel	10.2374	8.8082	0.0947
		Prewitt	10.2367	8.8221	0.0947
		Roberts	10.4018	9.7879	0.0912

feeble edge values. Additionally, a higher PSNR demonstrates that the reproduction is better from compressed images. In any case, sometimes like edge detection PSNR ought to be lesser to accomplish legitimate outcomes [3].

A measure of PSNR and SNR values is plotted for various edge detected methods for spine images in Fig. 7.3.

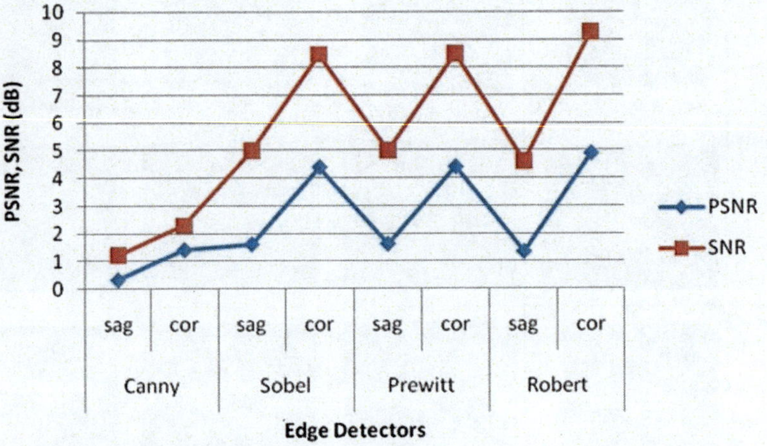

Fig. 7.3 PSNR and SNR of edge detectors

From Fig. 7.3, the Canny edge detected images give good PSNR values but the edges are not clear to diagnose the abnormality. Hence, we are in need of a good approach for segmentation.

7.7 Segmentation Method

Image segmentation is the process of partitioning homogeneous pixels in one group. Image segmentation is viewed as the underlying and standard activity to evaluate and understand spinal cord images. Edge detection is one of the basic methods of segmentation which detects the edges based on pixel intensity. But edge detection does not provide the fine details, so there is a necessity for novel approach of segmentation. One route for finding the ideal arrangement of edges is the thorough hunt technique. The complete search technique dependent on the Otsu model is straightforward, yet it has an impediment that it is computationally costly. In Otsu's strategy, best possible threshold points are liable for augmentation of difference between various groups. The issue of thresholding turns out to be progressively entangled if the threshold points increases. At the point when the images are extensively huge, utilizing a rapid and productive technique is exceptionally best. Particle swarm optimization (PSO) is an amazing enhancement strategy propelled by the social conduct of creatures living or moving in swarm and is significantly faster than exhaustive methods.

Algorithm

May there be L intensity levels in each RGB segment, and these levels are in the range

$\{0.1.2 \ldots \ldots L-1\}$, [12] then one can characterize

$$p_i^c = \frac{h_i^c}{N}, \sum_{c-\text{R,G,B}}^{N} i = 1, \; p_i^c = 1 \tag{7.18}$$

where I speaks to a particular intensity levels, i.e., $0 \ldots \ldots L-1$ and C speaks to the segments of the image, for example, {R, G. B}, N speaks to the overall number of pixels in the image, where h_i^c is the histogram for every segment C, which can be standardized and viewed as probability distribution p_i^c. The total mean (combined mean) of every part of the picture can be effectively determined as

$$\mu_C^T = \sum_{i=1}^{N} i P_i^C \text{ where } C = \text{R, G, B} \tag{7.19}$$

If we are going for a two-level threshold, then the segmentation is obtained by

$$F^C(x, y) = \begin{vmatrix} 0, f^c(x, y) \le t^c \\ L, f^c(x, y) > t^c \end{vmatrix} \text{ where } C = \text{R, G, B} \tag{7.20}$$

Presently, we can dichotomize the pixels of the given image into two classes isolating object from surroundings utilizing a limit level t^c. The probabilities of occurrence of two classes are given by

$$W_1^C = \sum_{i=1}^{t^c} p_i^c \text{ where } C = \{\text{R, G, B}\} \tag{7.21}$$

$$W_2^C = \sum_{i=t^C+1}^{N} p_i^c \text{ where } C = \text{R, G, B} \tag{7.22}$$

The least complex and computationally progressively effective technique for acquiring the best possible limit is the one that maximizes the between class variance characterized by:

$$\sigma_{\text{B}}^{2c} = W_1^C \left(\mu_1^C - \mu_t^C\right)^2 + W_2^C \left(\mu_2^C - \mu_t^C\right)^2 \tag{7.23}$$

where μ_1^C and μ_2^C are the mean of each class.

Our advancement issue is to look for the thresholds that maximizes the three target functions (i.e., fitness function) of each RGB segment commonly characterized as

$$\phi^C = \max_{1-t^C<L} \sigma_{\text{B}}^{2C}(t^C) \text{ where } C = \{\text{R, G, B}\} \tag{7.24}$$

The two-level thresholding can be reached out to n level thresholding likewise in which $n - 1$ threshold levels are required. So the pixels of the image will be separated into n classes, and between class difference can be characterized by

$$\sigma_{\text{B}}^{2C} = \sum_{j=1}^{n} W_j^C \left(\mu_j^C - \mu_t^C\right)^2 \text{ where } C = \{\text{R, G, B}\} \tag{7.25}$$

where j indicates a specific class.

Objective function for n classes can be written as

$$\phi^C = \max_{1<t_1^C<\ldots<t_{n-1}^C<L} \sigma_{\text{B}}^{2C}(t^C) \tag{7.26}$$

Processing this enhancement issue includes a progressively more computational exertion as the quantity of limit levels increment. As an answer for this, we consider PSO method.

The velocity v_n of the swarm is given by

$$v_n = wv_n + \rho_1 r_1(\overline{g}_n - x_n) + \rho_2 r_2(\overline{x}_n - x_n) \tag{7.27}$$

where ρ_1 and ρ_2 are constant integer values cognitive, and social components r_1 and r_2 are standardized random numbers between [0, 1]. The position x_n of the particle n is modified utilizing the following formula:

$$x_n = x_n + v_n \tag{7.28}$$

The particles in the PSO are assessed for the fitness function, which is characterized as the between class variance as meant in Eq. (7.23) of the image intensity distributions and tries to find the maximum of that threshold after many iterations. PSO uses this optimized value for segmentation of spine images.

PSO algorithm is applied for enhanced and bilateral-filtered image for four different patient's MRI images taken in sagittal and coronal view keeping the number of swarms $N = 150$. Table 7.7 shows the image after segmentation using PSO

Table 7.7 Segmentation using PSO

	Sagittal view		Coronal view	
	Image 1	Image 2	Image 3	Image 4
Original image	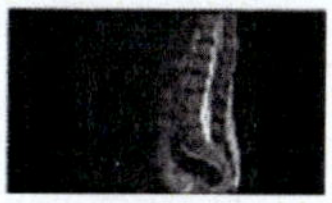	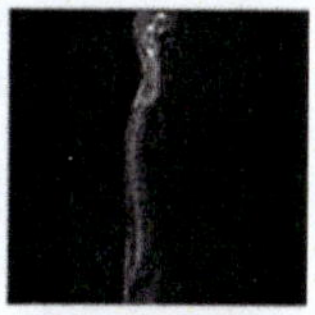	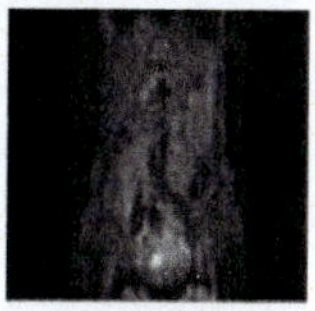	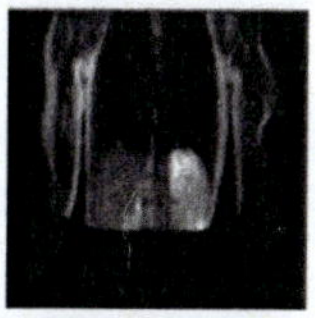
CLAHE image	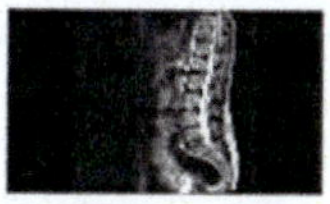	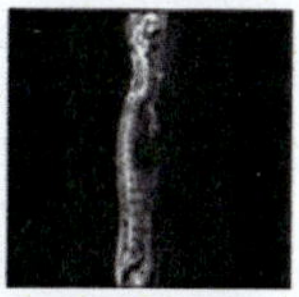	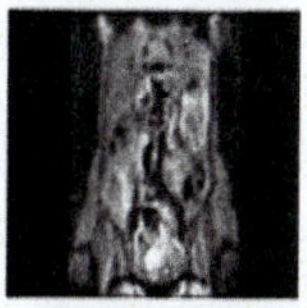	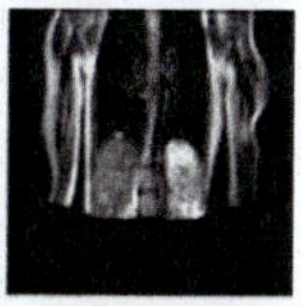
Bilateral-filtered image	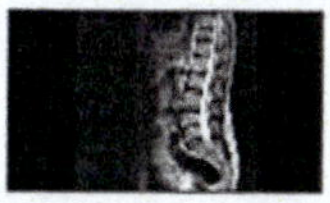	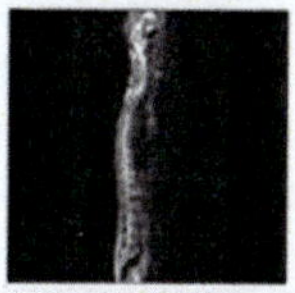	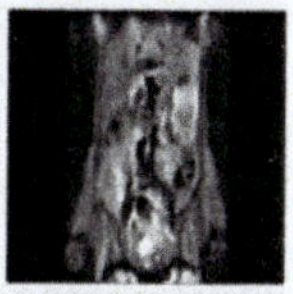	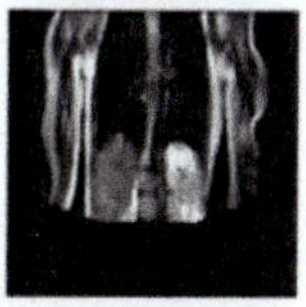
PSO segmented image	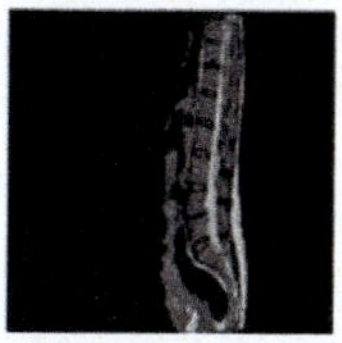	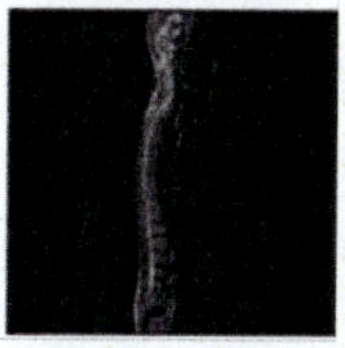	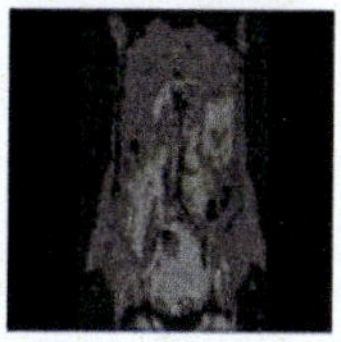	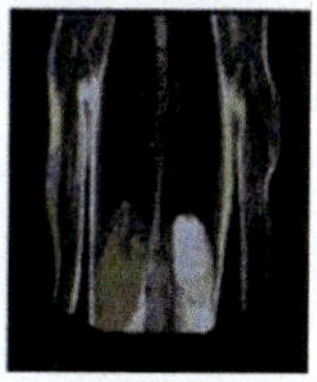

(continued)

Table 7.7 (continued)

	Sagittal view		Coronal view	
	Image 1	Image 2	Image 3	Image 4
Binary image	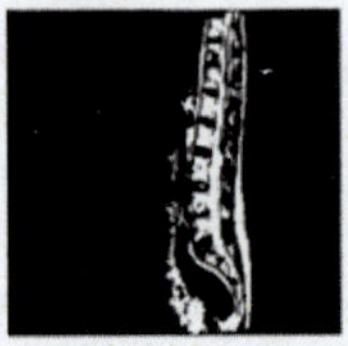	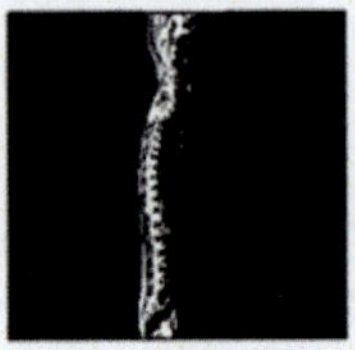		

Table 7.8 Dice coefficients for PSO method

Dice coefficient	Image 1	Image 2	Image 3	Image 4
PSO	0.4108	0.17330	0.4526	0.4510

method the Kappa index KI (also called dice coefficient) measures the quality of the segmentation [12]

$$\mathrm{KI} = 2 \times \frac{\mathrm{TP}}{2 \times \mathrm{TP} + \mathrm{FP} + \mathrm{FN}} \tag{7.29}$$

where higher value of Kappa index indicates better segmentation and where true positive is specified as TP, false positive is specified as FP and false negative is specified as FN.

After analysis of various edge detection algorithm for human vertebral spine image, it is found that edge information is not significant. Hence, the original image as shown in Table 7.7 is applied for denoising, and PSO algorithm is implemented for segmentation.

The Kappa index is calculated for PSO with $N = 150$ (no. of swarms) and n.o of iterations are 500.

The dice similarity coefficient (DSC) was utilized as a statistical evaluation metric to assess the presentation of both the reproducibility of manual segmentation and the spatial overlap exactness of programmed probabilistic partial segmentation of MR images. At the point when Kappa index = 1, the two segmentations are indistinguishable and 0 when no grouped pixel matches with exact segmentation. From Table 7.8, the DSC esteems are over zero which shows some amazing segmentation results.

7.8 Results

Experiments were conducted in the lumbar regions of patients MRI images. The analysis is done for two disk degeneration images and herniated images. As discussed in Table 7.2, the CLAHE method gives better enhancement than HE. The denoising is done using various methods, and from Table 7.4 bilateral filter gives good results. The CLAHE enhanced images are then filtered using bilateral filter, and PSO algorithm is applied for segmentation keeping no. of swarms = 150 and iterations = 500 (Table 7.9).

The dice coefficient is calculated for these above four abnormal spine images using PSO algorithm and tabulated in Table 7.10.

From Table 7.10, the dice coefficient is higher than zero which indicates the segmentation is better using our approach which aids in classification of abnormalities.

Table 7.9 Two abnormal images are analyzed using our approach

	Disk degeneration Image 1	Disk degeneration Image 2	Herniated Image 1	Herniated Image 2
Original image	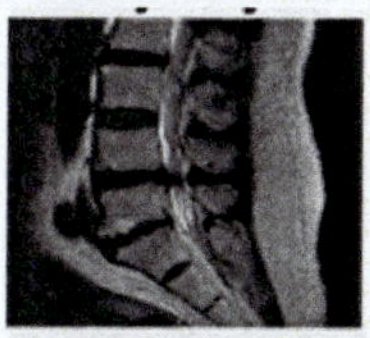	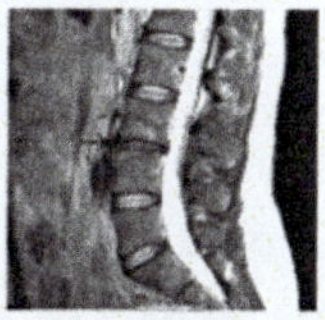	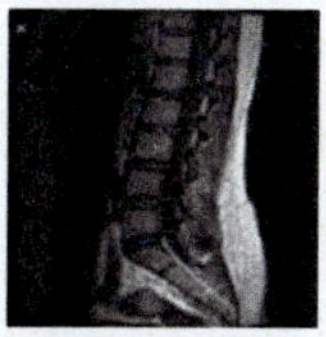	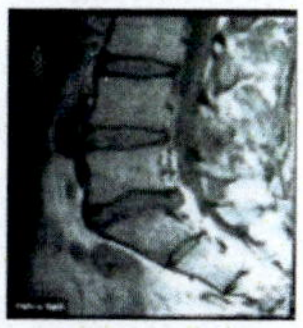
CLAHE enhanced image	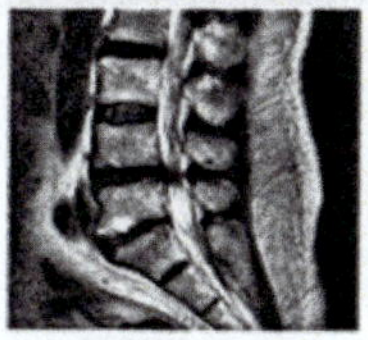	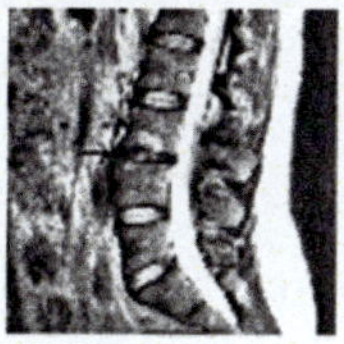	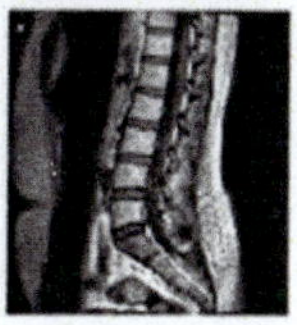	
Bilateral-filtered image	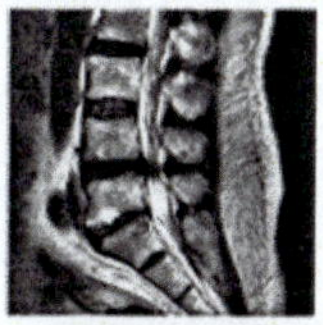	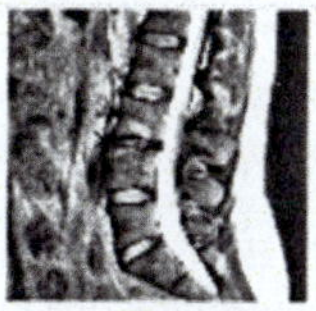	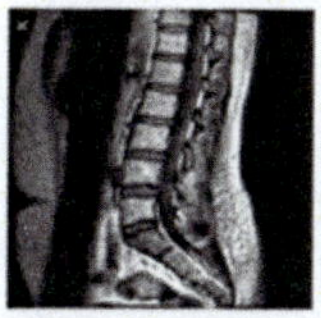	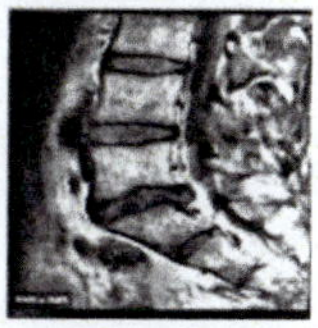
PSO method	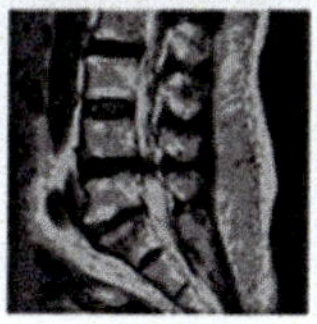	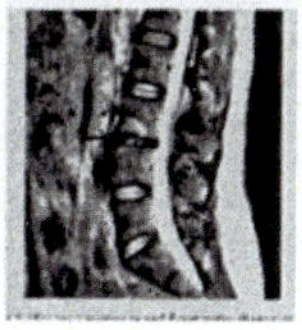	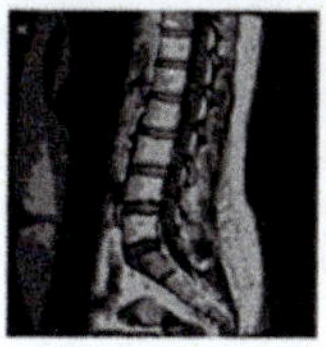	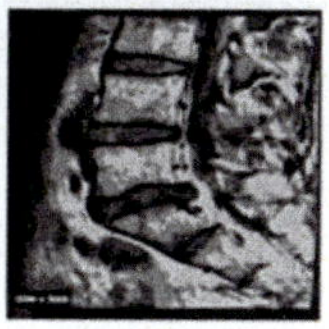

Table 7.10 Dice coefficient using PSO method for four abnormal images

Dice coefficient	Disk degeneration Image 1	Disk degeneration Image 2	Herniated Image 1	Herniated Image 2
PSO method	0.5560	0.345	0.5546	0.7725

7.9 Conclusion

In this paper, the preprocessing methods are compared for spine images. The enhancement of the image performed by HE and CLAHE method was performed and found CLAHE gives good quality of images which helps to extract the feature of an image. The various filtering methods are compared with their metrics and found the PSNR and SNR values are high and MSE is low for bilateral filter. Analysis of various edge detection techniques is applied and found the edges were not significantly determined. Hence, the original spine image is denoised and segmented using PSO method which out performs the basic edge detection methods. Trial results demonstrated that the present PSO approach is productive as far as dice coefficient. Here the segmentation of spine images is done using PSO method where this technique is applied to segment regions in satellite images. The abnormal region in the spine obtained after segmentation can be utilized to characterize the best pattern among the unhealthy images which helps in simple diagnosis.

References

1. J. Saenpaen, S. Arwatchananuku, N. Aunsri, A comparison of image enhancement methods for lumbar spine X-ray image, in *15th International Conference on Electrical Engineering/Electronics, Computer, Telecommunications and Information Technology* (IEEE, 2018)
2. E. Punarselvam, P. Suresh, R. Parthasarathy, Segmentation of CT scan lumbar spine image using median filter and canny edge detection algorithm. Int. J. Comput. Sci. Eng. (IJCSE) **5**, 806–814. ISSN: 0975-3397
3. P. Ghamisi, M.S. Couceiro, N.M.F. Ferreira, L. Kumar, *Use of Darwinian Particle Swarm Optimization Technique for the Segmentation of Remote Sensing Images. IGARSS 2012*. 978-1-4673-1159-5/12/$31.00 ©2012 (IEEE, 2012), pp. 4295–4298
4. I.A.M. Ikhsan, A. Hussain, M.A. Zulkifley, N.M. Tahir, A. Mustapha, An analysis of X-ray image enhancement methods for vertebral bone segmentation, in *2014 IEEE 10th International Colloquium on Signal Processing & its Applications (CSPA)*, Mar 2014 (IEEE, 2014), pp. 208–211
5. V.M. Georgieva, A.D. Mihaylova, P.P. Petrov, An application of dental X-ray image enhancement, in *2017 13th International Conference on Advanced Technologies, Systems and Services in Telecommunications (TELSIKS)*, Oct 2017 (IEEE, 2017), pp. 447–450
6. N.M. Sasi, V.K. Jayasree, Contrast limited adaptive histogram equalization for qualitative enhancement of myocardial perfusion. Sci. Res. 226 (2013)

7. S. Suhas, C.R. Venugopal, MRI image preprocessing and noise removal technique using linear and nonlinear filters, in *2017 International Conference on Electrical, Electronics, Communication, Computer and Optimization Techniques (ICEECCOT)*. 978-1-5386-2361-9/17/$31.00 ©2017 (IEEE, 2017)
8. E. Punarselvam, P. Suresh, Investigation on human lumbar spine MRI image using finite element method and soft computing techniques. Clust. Comput. Springer Science+Business Media, LLC, part of Springer Nature (2018). https://doi.org/10.1007/s10586-018-2019-0
9. J. Lee, S. Kim, Y.S. Kim, W.K. Chung, Automated segmentation of the lumbar pedicle in CT images for spinal fusion surgery. IEEE Trans. Biomed. Eng. **58**(7) (2011)
10. D. Poobathy, R. Manicka Chezian, Edge detection operators: peak signal to noise ratio based comparison. Int. J. Image Graph. Signal Process. **10**, 55–61 (2014). Published Online September 2014 in MECS (https://www.mecs-press.org/). https://doi.org/10.5815/ijigsp.2014.10.07
11. G.T. Shrivakshan, C. Chandrasekar, A comparison of various edge detection techniques used in image processing. IJCSI Int. J. Comput. Sci. Issues **9**(5), 269–276 (2012)
12. S.K. Hasane Ahammad, V. Rajesh, Md. Zia Ur Rahman, Fast and accurate feature extraction-based segmentation framework for spinal cord injury severity classification. IEEE Access **7** (2019). https://doi.org/10.1109/ACCESS.2019.2909583
13. S. Ait-Aoudia, E.-H. Guerrout, R. Mahiou, *Medical Image Segmentation Using Particle Swarm Optimisation* (IEEE, 2014). 1550-6037/14 $31.00 © 2014. https://doi.org/10.1109/IV.2014.68

Chapter 8
Smart Warehouse Management System

V. Sai Chitti Subrahmanyam, A. Vasantha Raman, S. L. Sharrik Krishna, I. Sitharthan, S. Shaheen Basha, Balasundaram Prabavathy, and S. Angel Deborah

Abstract A warehouse is the place where the food grains received from producers are stored before they are shipped to its respective fair price shops (FPS). Public distribution system (PDS) is a government-sponsored chain of shops entrusted with the work of distributing basic food and non-food commodities to the needy sections of the society at very cheap prices. Whenever there is a need or demand for food grains, the grains are transported from the warehouse to the FPS for public distribution purposes. Current PDS has limitations such as wastage of grains, fake beneficiaries, bogus cards, inferior quality food grains, corruption, non-transparency, non-accountability, inaccurate quantity of goods, long waiting time, low processing speed and material theft. In this context, wastage of grains indicates the loss of grains from the warehouse to the FPS. The distribution of inferior quality of food grains causes serious health issues for consumers which can be fatal. If the above-mentioned problems are addressed, more consumers can be served with better quality food. Currently, there is a need for manual intervention to maintain stocks of grain sacks and to check their vulnerability at warehouse. However, this can further be extended to improve the transparency in quality and transportation of food grains. Hence, this paper proposes SmartWMS which includes sensors with microcontroller at the warehouse to detect the moisture content in the grain sacks. Based on the moisture content, the sacks are grouped into sacks with moisture below threshold (MBT sacks) and sacks with moisture above threshold (MAT sacks). An Android application is developed to automate the transportation of MAT sacks to the FPS.

Keywords Public distribution system · Fair price shops · Smart warehouse management system · Food Corporation of India · Android application

---

V. Sai Chitti Subrahmanyam (✉) · A. Vasantha Raman · S. L. Sharrik Krishna · I. Sitharthan · S. Shaheen Basha · B. Prabavathy · S. Angel Deborah
SSN College of Engineering, Chennai, Tamil Nadu, India

B. Prabavathy
e-mail: prabavathyb@ssn.edu.in

S. Angel Deborah
e-mail: angeldeborahs@ssn.edu.in

R. Seyezhai et al. (eds.), *Recent Trends in Renewable Energy Sources and Power Conversion*, Springer Proceedings in Energy,
https://doi.org/10.1007/978-981-16-0669-4_8

8.1 Introduction

According to the report of Food and Agriculture Organization (FAO) stated in [1], 194.4 million people are undernourished in India. From these statistics, 14.5% of the total population is undernourished, in which 51.4% of women are anemic and 37.9% of the children are undergrown. Malnourished children have a higher risk of death from common childhood illnesses such as diarrhea, pneumonia and malaria. The Global Hunger Index ranks India at 103 out of 119 countries on the basis of two leading indicators, namely undergrowth in children who are less than 5 years and death rate of children who are below 5 years.

The root cause of the above-mentioned problem can be traced to wastage of food grains. It is estimated that nearly one third of the food produced for human consumption every year gets lost or wasted. Forty percent of the fruits, vegetables and 30% of cereals that are produced do not reach the consumer markets due to inefficient supply chain management. According to the United Nations Development Program, about 21 million tons of wheat, the primary grain, is wasted in India, which accounts for INR 5000 crores every year. This alarming situation can be addressed by curbing down the factors that cause the wastage of food grains. Agricultural industry has always experienced losses due to spoilage of harvested grain in long-term storage, as cited in [2].

Post-harvest loss can be defined as the loss from the stage of harvesting to the stage of consumption which occurs as a result of qualitative loss, quantitative loss and the food waste (by the consumers) altogether. Poorly controlled moisture and temperature conditions in the storage area lead to a chain reaction of spoilage of grains by mold and insect growth, as stated in [3]. Indian farmers incur Rs. 92,651 crore per year in post-harvest losses, the primary causes of which are poor storage and transportation facilities.

It is the responsibility of the government of a country to ensure that their people have access to basic diet everyday despite all the above-mentioned problems. The World Food Summit of 1996 defined food security as existing, "when all people at all times have access to sufficient, safe, nutritious food to maintain a healthy and active life". In India, the public distribution system was invented by British government in 1939, and it was improved and modified over the time by Indian government to ensure food security. Now, PDS is under Food Corporation of India (FCI) which is the only government agency entrusted with movement of food grains from the procuring states to consuming states through a network of storage infrastructure owned or hired by FCI in the whole of India, as stated in [4]. It chiefly sells the grains—wheat and rice, at a price lower than the market price. In general, these grains are supplied from warehouse to fair price shops (FPS).

Warehouse management is to ensure an efficient way of organizing the process of flow of grains from the point they are acquired to the time they reach the distribution centers. The current warehouse management methodology involves a series of manual work in the process of management. These systems are labor intensive. Further, the manual steps involved often lead to errors in the system's results. Hence, it is highly important to improve the efficiency of the system to ensure that the food grains produced by hard work of farmers are not wasted in order to serve the people in a better way.

Moisture value plays a vital role in determining the quality of food grains that are stored, as described in [5]. Hence, capturing the moisture value is very essential. Currently in FCI, the readings of the moisture meter are noted manually and are fed into the server. Since there is a human intervention, there could be chances that the recorded readings could be erroneous. The work addressed in [6] has developed a network module to transfer the data from moisture meter to the server. This system is limited to the automation of detection of goodness of the grains in the warehouse itself. However, grains have to be quickly moved to the nearest FPS to reach the public. Hence, the objective of this paper is to propose a smart warehouse management system (SmartWMS) to process the collected moisture data in order to efficiently transport the food grains.

8.2 Existing Work

Public distribution system (PDS) is a mechanism to enhance the food security in India. It faces challenges like leakage of grains, fake beneficiaries, bogus cards, inferior quality food grains, corruption, non-transparency and non-accountability. The major challenge is the presence of ghost beneficiaries. This is due to inaccuracy in the identification of beneficiaries. The absence of an effective identification mechanism was one of the biggest barriers for the poor to access the services.

In this context, Hitaswi and Chandrasekaran [7] have proposed a solution which utilizes the Aadhar number. In this paper, there exists an empirical study on the unique identification mechanism to identify the beneficiaries in PDS and agent-based social simulation of PDS. If the identification mechanism is related to Aadhar, the major problem of bogus and duplicate cards will be removed. This biometric verification helps to reduce the leakage of commodities from fair price shops. Further, they also proposed a simulation which helps to connect all the agents such as farmers, FCI, state government, fair price shops and the beneficiary involved in the PDS using the agent modeling tool, namely, Net logo.

In order to address low processing speed, Priya and Nikumbh [8] have proposed E-PDS which replaces existing ration cards with smart cards. Main objective of the designed system is to replace manual work with the atomization of ration shops to have transparency in PDS. A smart card consists of details of the family, type of card, type of validity (to name a few) with respect to each consumer. In this E-PDS, a consumer needs to scan the smart card, and if it is validated, he has to enter the materials and quantity to be purchased. Then, the materials are automatically dispatched without human interpretation. Thus, the system aims at avoiding corruption and ensures transparency at each level. The proposed PDS system has its point of sale (POS) outlets centered around an e-box 3310 MSJK device which runs an OS customized in Windows CE. This processor communicates with three different centralized databases to access different data associated with the customer. This communication is triggered by a unique identification (UID) smart card. This work proposed an automation of the provision of ration. At first, UID and fingerprint details of the consumer will be passed onto the processor for verification. Weighing machine measures the exact weight of goods being dispersed based on the consumer's request.

Food grain contamination occurs due to inappropriate ambient conditions when the grains are being transported and stored. In this context, Srivastava and Gulati [6] have proposed an IoT framework to analyze the ambient conditions under which the food grains are stored and transported. In this solution, heterogeneous sensors for various domains are employed for sensing the condition of food. It senses the temperature, moisture and light parameters of the surrounding environment. These values are then accessed by the user, and if these values are greater than the threshold values, the user is notified, and appropriate steps are taken to store under ambient conditions. Thus, integration of the sensors with remote Web servers and a real-time user notification mechanism ensures in protecting food grains from being contaminated and wasted.

The flow of the food grains through the supply chain can be divided into three major parts, from farmers to warehouse, from warehouse to FPS and from FPS to end consumers. The previously proposed systems emphasize on improvising the first part of the supply chain or the second part exclusively. The previously proposed systems for warehouses are limited to monitoring of the storage of grains under a single roof. The newly proposed system explores the area of improvising the quality of grains in their storage time, minimizing the possibility of loss of grains while adding to the transparency and accountability to the maintenance of stocks and thus addressing two major problems, namely leakage of grains in the supply chain and wastage of food grains in the public distribution system.

8.3 Proposed System

The proposed SmartWMS shown in Fig. 8.1 aims to automate the functionalities of the warehouse and transportation of grains to the fair price shops (FPS). It works as follows.

- Grain sacks in the warehouse are associated with barcodes.
- Moisture sensor is used to detect the moisture content of the grain sacks.
- Barcode scanner scans the barcode of the grain sacks.
- NodeMCU retrieves the sensed moisture value of the grain sacks along with their barcodes.
- NodeMCU pushes the retrieved data into Firebase using its Wi-Fi module.
- Firebase stores the sensed moisture values for every grain sack.
- An Android application is developed to manage the activities of the warehouse. The app is named as WM-App.

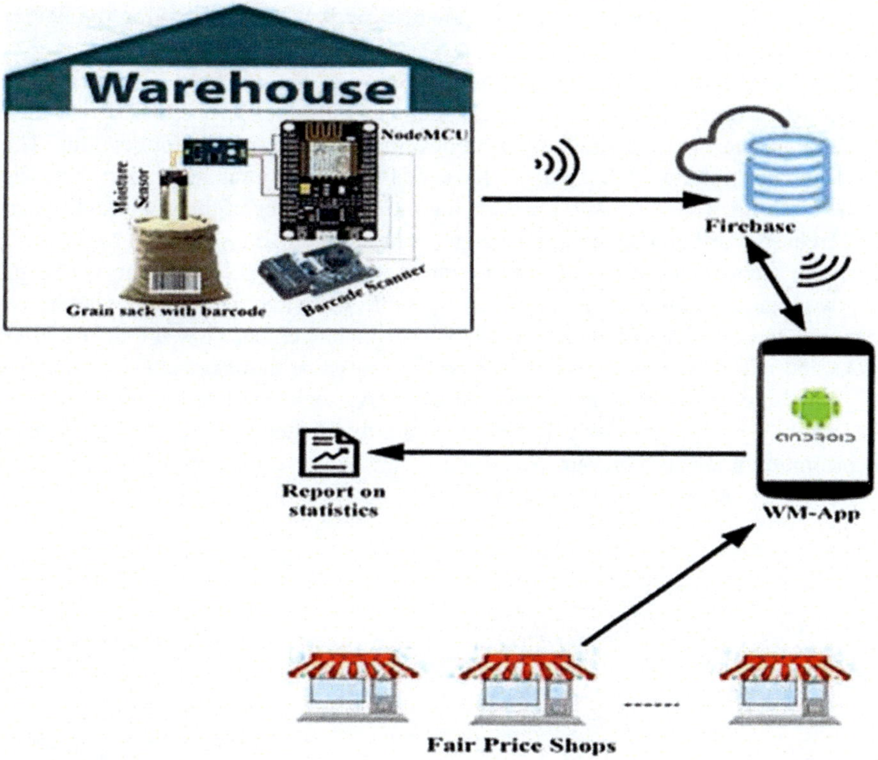

Fig. 8.1 Design of SmartWMS

- WM-App retrieves the sensed values from the Firebase and categorizes the grain sacks into MBT and MAT sacks.
- WM-App automates the transportation of the MAT sacks to the nearest FPS and generates the route map.

The warehouse has a number of grain sacks that are to be monitored. Each grain sack has a barcode associated with it, which helps to identify the sack distinctly. SmartWMS uniquely identifies the grain sacks with the help of a barcode scanner and also reads the moisture values associated with that grain sack through a moisture sensor. The barcode scanner uses serial communication to transmit the identified barcode to the NodeMCU.

NodeMCU microcontroller is an open source software and hardware development environment that helps to read values from the hardware component. It has 16 general purpose input/output pins (GPIO). Also known as ESP8266 module, it houses a Wi-Fi module that is supported under 802.11n standards with 2.4 Hz frequency. The moisture sensor uses electrical resistance between the probes to sense moisture level and can be adjusted with the potentiometer. Both the barcode scanner and the moisture sensor are under the control of a single push button, which triggers the central housed

NodeMCU. The NodeMCU is in turn powered by 9 V battery, making the device handy.

NodeMCU retrieves the values from the sensors. The received values are then transmitted to a remote cloud database, namely Firebase. Firebase is a real-time database hosted in cloud which helps in developing collaborative applications. The Wi-Fi module of NodeMCU is utilized to send the moisture readings and the barcode to the Firebase. The WM-App retrieves the moisture values of the grain sacks from the Firebase. Further, it categorizes the grain sacks into MBT sacks and MAT sacks based on their moisture values. The count of MAT sacks is found. The WM-App has two login interfaces designed for two persons in charge, namely the warehouse manager and the FPS manager. The warehouse manager can view the list of MAT sacks and initiate their transportation to FPS. Fair price shops are chosen such that they have enough space to accommodate the MAT sacks and are also proximate to the warehouse. Once received by FPS, they are distributed to people before further deterioration happens. The functionalities of the WM-App are explained briefly in the following use case diagrams shown in Figs. 8.2 and 8.3.

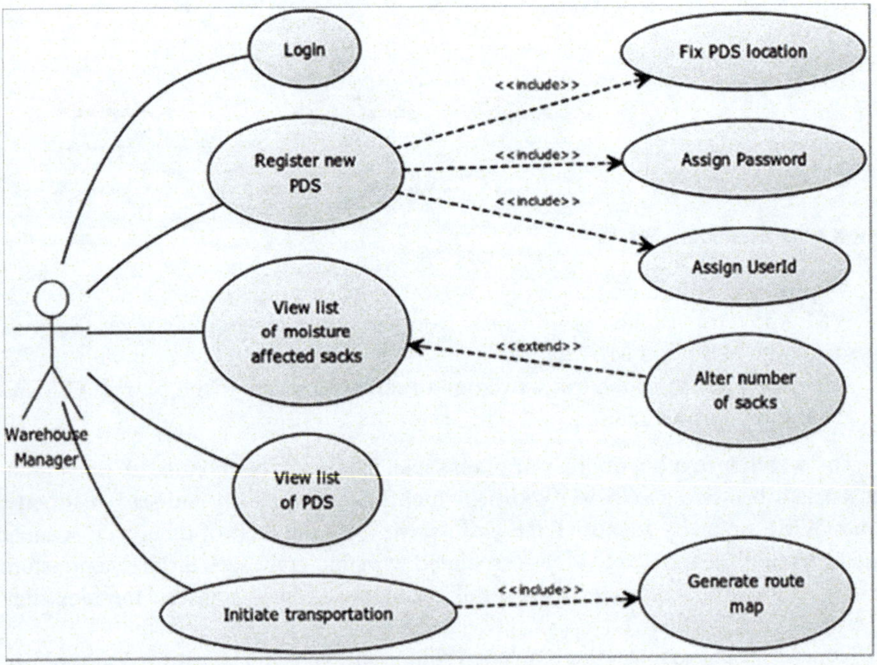

Fig. 8.2 Use case description for warehouse manager

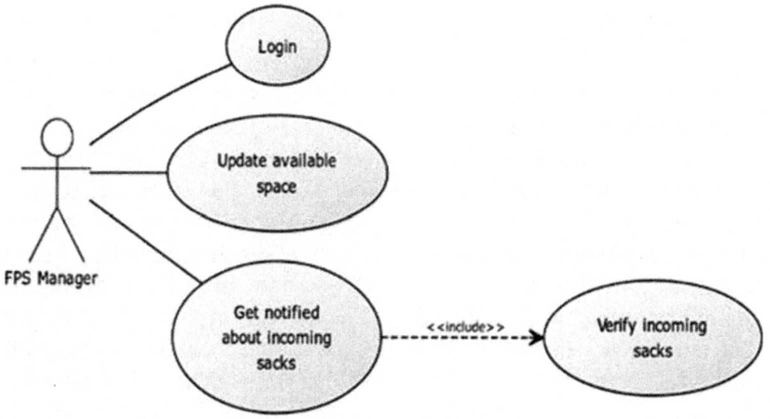

Fig. 8.3 Use case description for FPS manager

8.3.1 Use Case Description for Warehouse Manager

The warehouse manager is the controller of the warehouse's information. He can login to the WM-App to access the moisture readings of the grain sacks. When there is a need, warehouse manager registers new FPS near the warehouse, and he/she is the concerned person to register the FPS in the WM-App. The registration process includes setting up of a user ID, a password and specifying the location of that corresponding FPS. When there is a rise in the moisture levels of the food grains, the warehouse manager will be able to identify the MAT sacks and update the number of MAT sacks in the Firebase. He can also view the list of FPS nearer to the warehouse, and the number of sacks the FPS can accommodate from which the warehouse manager decides where to transport the MAT sacks on time. The transportation is initiated by the warehouse manager, and the route map is generated for the selected FPS location.

8.3.2 Use Case Description for FPS Manager

The FPS manager is the controller of his FPS's activities. The FPS owner can login to the FPS's login of the WM-App to update their available storage spaces regularly. He/she will be able to update the number of sacks that can be accommodated in their FPS frequently. Based on their available spaces, warehouse manager decides to send MAT sacks to the nearest FPS. The FPS owner gets notified by the number of incoming sacks. Upon the arrival of the sacks, the FPS manager verifies the number of incoming sacks and acknowledges the same to the warehouse manager.

8.4 Implementation

The prototype of SmartWMS shown in Fig. 8.4 involves four phases, namely sensing, acquiring, transmitting and processing. In the sensing phase, the moisture sensor has been utilized to read the moisture values of a particular sack of food grains, which is uniquely identified by a barcode that is read using a barcode scanner. The moisture sensor contains probes which have to be inserted into the food grains, and it utilizes the electrical resistance property to measure the moisture of the grains. The barcode scanner is used to detect the barcode of a particular sack of food grains for which the moisture reading is to be taken. The barcode scanner uses infrared light to scan the barcode on click of a button which also activates the moisture sensor to capture readings simultaneously. The barcode scanner is set to serial communication mode with baud rate of 9600 compatible with NodeMCU. The barcode scanner uses TTL-232 interface for serial communication and operates in the range of DC 4.2–9 V. The barcode scanner is also mounted with white flash light for operation in dark environments. The read time for one scan is approximately three seconds and requires a read interval of minimum one second.

The Arduino IDE has been utilized to write the code to acquire the data. This code is burnt onto the on-chip memory of NodeMCU. Once burnt, the NodeMCU can be powered by an external power source (5–9 V), making it a wireless handheld device. The process is initiated on a button press that is connected physically to a NodeMCU board via port 3. The board is programmed in a way so as to power up the moisture sensor connected to the board via port 5. The barcode scanner is connected to NodeMCU via port 7 and port 8 to felicitate serial communication as shown in Fig. 8.5.

In the transmission phase, Firebase has been utilized to store the moisture data. Firebase has been chosen in order to retrieve the data from any place. In order to implement this, it is necessary to create an account in Firebase. With that account,

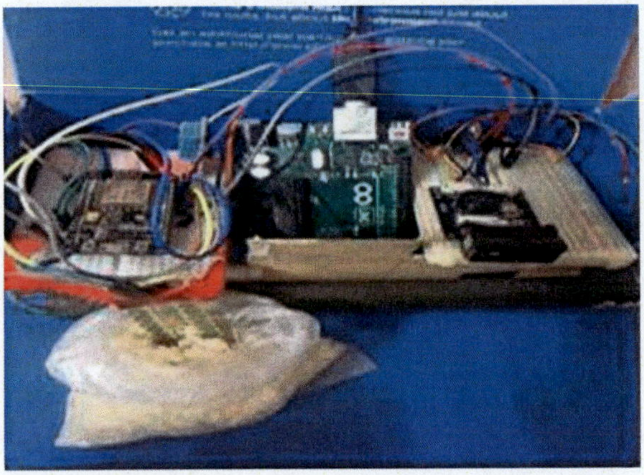

Fig. 8.4 Prototype of SmartWMS

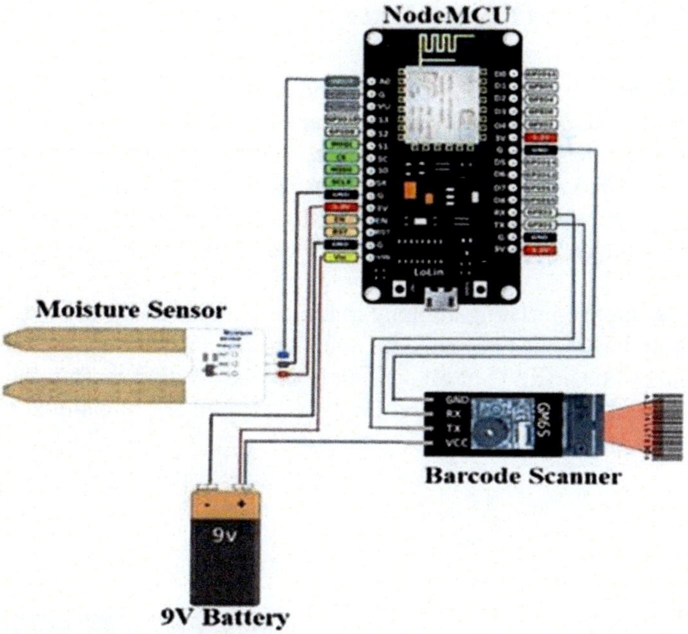

Fig. 8.5 Circuit diagram of prototype of SmartWMS

a unique reference id is created. This reference id of the Firebase account is used in the program that is burnt in the board to transmit any data to the Firebase.

In the processing phase, an Android application, namely WM-App, is developed to process the data. The WM-App is built using Android Studio. It has a collection of Java class files and corresponding XML layout files. The WM-App can be run on all Android phones with Android version greater than v4.1.1 (Jelly Bean).

Workflow of WM-App can be described in two processes as given below.

8.4.1 Registration Process

The WM-App has two types of users, namely warehouse manager and FPS manager. Warehouse manager registers the FPS under his control. It can be done by providing a user ID, a password and the exact pinpoint location of new FPS in a map layout. This creates a marker object for the location pointed, from which the coordinates of the location and its distance from the warehouse are obtained. These coordinates contain the latitude and longitude of the area under consideration. Whenever a new FPS is registered, a login ID and a password are generated for the new user to become a part of the system. His record gets updated in the Firebase with fields of password, location.

8.4.2 Demand and Supply Process

FPS manager has to update the stocks used and its available storage space in the WM-App on a regular basis. This change is reflected on Firebase. The WM-App processes the data from Firebase and categorizes them as MBT and MAT sacks based on the threshold moisture value. Once the warehouse manager logs into WM-App, he can view the real-time organized data of grain sacks, as shown in Fig. 8.6 and can initiate the process of transfer of MAT sacks to the nearest FPS from the list provided in the WM-App.

The maps application programming interface (API) is used to find the distances between warehouse and FPS, possible routes to the FPS and their corresponding expected arrival time (EAT) of all possible routes to every FPS. This data given by maps API and sack space data from Firebase are used to optimally select FPS as described in Algorithm 1.

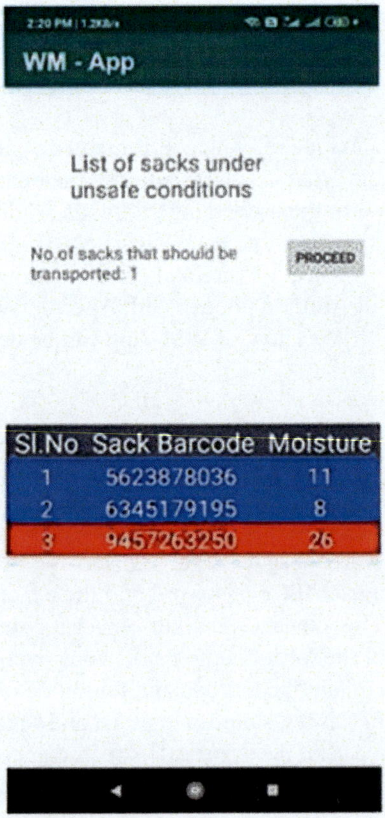

Fig. 8.6 WM-App of SmartWMS

Algorithm 1: Supply and Demand Algorithm

Input. Warehouse object, list of FPS objects
Output. One or more FPS

Warehouse object data members:

- location //warehouse location obtained while registration
- num_MAT //count of MAT sacks
- num_MBT //count of MBT sacks

FPS object data members:

- location //location of FPS obtained while registration
- route //route from Warehouse to FPS
- distance //distance from warehouse to FPS
- EAT //Estimated arrival time (warehouse to FPS)
- demand //demand of each FPS

```
let wh = warehouse
let fps_list = list of FPS under service of wh

Retrieve barcode and moisture values from Firebase Cate-
gorize sacks as MAT and MBT based on threshold moisture
value.

for f in fps_list
   //maps_api returns routes between given two locations
  routes[ ] = maps_api(f.location,wh.location)
  optimal_route = routes[0]
  for r in routes
    if (r.EAT < optimal_route.EAT )
      optimal_route = r
  f.route = optimal_route
  f.distance = optimal_route.distance
  f.EAT = optimal_route.EAT

sort fps_list based on distance

total_demand = 0
k = 0
//find total demand of all FPS
for f in fps_list
  total_demand = total_demand + f.demand
```

```
//while total demand > 0, supply from MAT sacks
while (total_demand > 0 and wh.num_MAT > 0)
  demand_fps = fps_list[k++].demand
//exhaust MAT sacks and then supply from MBT sacks
if(demand_fps > wh.num_MAT)
  new_demand_fps = demand_fps - wh.num_MAT
  fps_list[k-1].demand = new_demand_fps
  total_demand = total_demand - wh.num_MAT
  wh.num_MAT = 0
  k--

//if demand is less than MAT sacks, supply from MAT sacks
  wh.num_MAT = wh.num_MAT - demand_fps
  total_demand = total_demand - demand_fps

//if total demand still exists, supply from MBT sacks
while (total_demand > 0)
    demand_fps = fps_list[k++].demand
    wh.num_MBT = wh.num_MBT - demand_fps
      total_demand = total_demand - demand_fps
```

8.5 Features of Proposed SmartWMS

The proposed SmartWMS is an end-to-end solution to keep track of quality and transportation of food grains from warehouse to FPS. The existing works focus only on monitoring food grains either at the warehouse or at the FPS. However, SmartWMS connects the warehouse as well as the FPS by transporting the MAT sacks from the warehouse to the nearest FPS on time. The following are the features of SmartWMS.

8.5.1 Automation of Moisture Reading

In the existing system, a person in the warehouse notes down the moisture values of the grain sacks and enters them into the database manually. It is a time-consuming process. During this process, there are lots of other issues faced too. He/she might not check the moisture values periodically in each sack. There is no instantaneous updating of the moisture values along with sack's unique ID. Manual errors may take place such as entering erroneous values by the person into the database. Due to this, there is a need to monitor the moisture levels in each sack using some unique identification. The existing system has identification of each sack through numbered sacks using chalk like material, but they are not secured enough. Due to such manual errors, there are tons of wastage of food grains in the country as specified by the report of FCI.

Whereas in the proposed SmartWMS, the drawbacks due to the manual intervention at the time of reading and updating the moisture values into the database are prevented by the following approach. Each sack consists of a unique barcode identification label on it. The reading person has to just insert the moisture sensor into the sack to measure its moisture value and scan the barcode label on the sack using the barcode scanner. The Wi-Fi module of the NodeMCU is used to transfer the sensed moisture values along with barcode ID from the warehouse to the Firebase directly for further processing by the WM-App. Because of its low-cost platform to perform in IoT, NodeMCU is used in the system. Hence, SmartWMS automates the moisture readings and reduces human errors from updating the erroneous values into the system.

8.5.2 Remote Access of Data

In existing systems, the sensed moisture values are manually updated by the warehouse manager into the server of the warehouse at fixed intervals of time. Hence, the warehouse manager can access data only within the warehouse and not from anywhere else as each warehouse has centralized recorded data in the server. So, in case of any accident or damage happens to the server, the data is permanently lost. Also, the warehouse manager has no access to the details of nearest FPS present.

Whereas in the proposed SmartWMS, the sensed moisture values are directly updated into the cloud database. The use of cloud database guarantees no loss of data due to maintenance of redundant copies of data in the cloud and thus ensures more security of data. Hence, the warehouse manager can access data from any place at frequent intervals of time through the WM-App. The warehouse manager can also view the details of newly registered FPS and the amount of space available in each nearest FPS to efficiently transport sacks on demand.

8.5.3 Flexibility

In the existing system, only a fixed amount of sacks is allocated to each FPS. The warehouse manager is not aware of the exact amount of space left in FPS to accommodate the grain sacks. Also, the location of FPS may change sometimes within their locality, and this change may not be noted by the warehouse manager dynamically. Adding onto this, whenever there is a new FPS established in some area near the warehouse, it is not noted and used immediately.

Whereas in the proposed SmartWMS, the space left in each FPS is updated frequently in the Firebase which can be accessed by the warehouse manager at any point of time. Except the WM-App, there is no need of other communications between the warehouse and the FPS to know the changing amount of spaces left in the FPS at frequent instances of time after distributing grains to the customers. In addition to this, when there is a change in the location of an FPS, the warehouse manager is instantly updated with the new location's coordinates of the FPS through the WM-App. Also, whenever there is a new FPS that is established and comes under the service of WM-App, the registration of the FPS is a quick process in the WM-App.

8.5.4 Efficient Transportation

In existing systems, once the MAT sacks are identified, the warehouse manager is just notified regarding their condition. Also, they are transporting on regular basis but not based on the demand in FPS.

Whereas in SmartWMS, once the MAT sacks are identified, it instantaneously provides the warehouse manager with an optimal way to transport them. The space availability in each FPS has been updated regularly in the WM-App by each FPS manager, and WM-App utilizes this data to make an optimized decision to transport the MAT sacks to appropriate FPS. The WM-App uses the Google Maps API to give the efficient routes for transportation based on distance and estimated arrival time.

8.5.5 Transparency in Monitoring

In existing systems, the warehouse manager does not know the status of transported sacks of food grains. Once the food grains are transported, the warehouse manager cannot monitor the sacks and keep track of them.

Whereas in proposed SmartWMS, the warehouse manager can track the status of all transported sacks of food grains, i.e., whether they have reached their destination safely or not. If sacks are found to be missing, the FPS manager can identify the missing sacks with its barcode and can hold the transport people accountable for the same. The warehouse manager is also notified if the transported sacks are found to be missing. Thus, the grains sacks are secured and safely transported.

8.6 Conclusion

The proposed SmartWMS provides an end-to-end solution to keep track of the quality and transportation of food grains from warehouse to FPS. SmartWMS automates the moisture readings of uniquely identified sacks of food grains directly into the Firebase with the help of a moisture sensor and a barcode scanner, thereby reducing human errors from updating erroneous values into the system. Since data is recorded in the cloud database, redundant copies of data are present and can be accessed by the warehouse manager from anywhere through the WM-App. The WM-App processes the sensed moisture values and categorizes them into MAT and MBT sacks based on the threshold moisture values. Then, the MAT sacks are transported efficiently to the nearest FPS based on the availability of sacks.

Hence, SmartWMS provides an efficient transportation mechanism for distribution of sacks on demand and also ensures that the demand–supply chain is maintained between the FPS and the warehouse. Further, it improves the transparency in monitoring and managing the flow of food grains in the supply. In this way, SmartWMS prevents wastage of food grains during their storage period in the warehouse.

8.6.1 Limitations

- The process is dependent on rate of decay of grains, and adverse conditions might still affect grains.
- The system does not state ways to prevent malpractices existing in the food supply chain.
- Wastage of food grains due to pest attacks cannot be prevented.

8.6.2 Scope for Future Works

- 1D barcodes can only hold a maximum of 85 characters, while 2D barcodes can store over 7000 characters, thus allowing to transmit more information. The future system can move to 2D barcodes, and warehouses would be able to convey much more complex information, like serial numbers, grain names, etc., all without the need for any additional scanning.
- Aeration circuits could be introduced into warehouses that could optimize the entire room of required temperature and moisture.
- Fire sensor, temperature sensor, humidity sensor and buzzer can be added to the warehouses to increase efficiency in risk management of grains.

Acknowledgements This project was funded by Sri Sivasubramaniya Nadar College of Engineering, Kalavakkam, Chennai, during the academic year 2018–2019.

References

1. The state of food security and nutrition, https://www.unicef.org/reports/state-of-food-security-and-nutrition-2019. Updated on 15 July 2019
2. 62,000 tons of food grain wasted by FCI, https://www.dnaindia.com/india/report-62000-tonne-foodgrain-wasted-by-fci-2446621. Updated on 22 May 2017
3. Food grains rot in India while millions live with empty stomachs, https://www.lowyinstitute.org/the-interpreter/food-grains-rot-india-while-millions-live-empty-stomachs. Updated on 16 Nov 2018
4. The public distribution system and food security in India, https://www.researchgate.net/publication/335628611_The_Public_Distribution_System_and_Food_Security_in_India. Updated on 3 Sept 2019
5. Moisture content in food grains storage, https://igmri.dfpd.gov.in/igmri/moisture. Updated on 27 Nov 2019
6. A. Srivastava, A. Gulati, iTrack: IoT framework for smart food monitoring system. Int. J. Comput. Appl. **148**(12) (2016)
7. N. Hitaswi, K. Chandrasekaran, Agent based social simulation model and unique identification based empirical model for public distribution system, in *2017 International Conference on Recent Advances in Electronics and Communication Technology (ICRAECT)*, 16 Mar 2017 (IEEE), pp. 324–328
8. B.G. Priya, S. Nikumbh, E-public distribution system using SMART card and GSM technology, in *2017 International Conference on Intelligent Sustainable Systems (ICISS)*, 7 Dec 2017 (IEEE), pp. 244–249

Chapter 9
Solar-Based Water Purification System for Urban Areas: A Case Study

Parthsarathi Mulay, Twinkle Rane, Namrata Kala, and Archana Thosar

Abstract Water is a necessity for the survival of all living beings. Even though 71% of the Earth's surface is water, there is still an acute shortage of potable water in many countries, as approximately 97.5% of Earth's water is saltwater in the oceans and only 2.5% is freshwater in groundwater, lakes, and rivers. Thus, access to clean drinking water is a major issue that needs to be tackled immediately and efficiently. Currently, existing technologies for water purification such as multistage flash, multiple effect, vapor compression, reverse osmosis (RO), ion exchange, electrodialysis, and capacitive deionization (CDI) require a lot of energy, leading to environmental pollution caused by the consumption of fossil fuels. It is also found that 70% of water supplied as input to these technologies is wasted, while methods such as RO remove important minerals, and hence, World Health Organization (WHO) standards do not deem this fit. The desalination of saline water using solar stills is an effective solution to overcome these problems. Solar energy is abundant, everlasting, economical and does not cause emission of greenhouse gases. Combining this worldwide availability and inexhaustible nature of solar energy with a desalination process, a water purification system is designed and operated. This paper proposes the design and application of a water purification system which will be beneficial for localities such as residential buildings and bungalows in densely populated cities where the rise in energy consumption has led to an increase in the carbon footprint, reduction in electric power supply, and higher energy costs along with the depletion of fossil fuels and other exhaustible energy sources. This design can be further modified and installed in arid regions and villages where grid electricity is not a feasible option to meet energy demands and water scarcity is prevalent.

Keywords Desalination · Solar stills · Mathematical model · Water survey

P. Mulay (✉) · T. Rane · N. Kala · A. Thosar
College of Engineering Pune, Pune, India
e-mail: parthsarathim16.elec@coep.ac.in

R. Seyezhai et al. (eds.), *Recent Trends in Renewable Energy Sources and Power Conversion*, Springer Proceedings in Energy,
https://doi.org/10.1007/978-981-16-0669-4_9

9.1 Introduction

The major challenges today are lack of freshwater sources for drinking and the depletion of the environment due to the burning of fossil fuels. With the rapid increase in industrialization and a demand to fulfill the needs of the fast-growing world population, the use of fossil fuels to provide energy has also increased. However, the emission of such fuels has led to environmental pollution and global warming. Hence, it has become the need of the hour to consider renewable energy resources to sustain the basic requirements of people. Solar desalination uses the most sustainable and clean energy source, i.e., solar energy to produce high-quality water. It is one of the most inexpensive methods to produce potable water [1] and can be installed in areas like India that receive a good amount of sunlight on a regular basis. Given the acute shortage of water in large and developing cities, people often face a water crisis. Technologies such as RO and ultraviolet filters are currently used which involve wastage of water along with other disadvantages such as higher consumption of energy supply. The abundant energy from the sun can be tapped to achieve a desalination-based water purification system. Eltawil et al. [2] have done comprehensive research on coupling various desalination technologies with renewable energy. It is normally considered that salinity below 500 ppm is acceptable for drinking water. Ali et al. [3] extensively studied different types of water desalination processes. Ayoub and Malaeb [4] have regarded simple solar stills as one of the most promising water purification techniques that can be effectively used to convert sea and brackish water into freshwater. Agrawal et al. [5] have studied the energy balance equations for the thermal modeling of the conventional solar still. Panchal and Thakkar [6] theoretically and experimentally studied the improved efficiency of a solar still coupled with evacuated tube collectors. The statistical parameters were calculated theoretically and showed close proximity to the experimental results.

In this paper, the mathematical model of the single basin solar still is simulated to determine the variation of the distillate obtained and to observe the variation of critical parameters on the output. A novelty in design by coupling single basin solar still with evacuated tubes is implemented, and the mathematical model for the same is simulated. Subsequently, a case study on the drinking water requirements in urban areas is presented to determine the feasibility of the proposed purification method.

9.2 Solar Stills

Different designs of solar stills varying from the most fundamental to the most advanced design have been studied and realized in the past. While the basic design has variations depending on its application and type, the actual design and idea did not change much since the stills built in the days of the Las Salinas, Chile, in 1872 [7]. A solar still works on the basic principle: As the incident radiation is transmitted through the glass cover, a part of the radiation gets absorbed by the cover while another part by the saline water. This increases the temperature of the water inside the basin. Due to the temperature gradient between the inner and outer sides of the basin, the water in it starts to evaporate, and the vapor rises up toward the glass cover by natural convection, where it cools down and condenses on the inner surface of the cover [8]. The condensate then flows naturally into the collection tank under the influence of gravity, and pure distillate is obtained as shown in Fig. 9.1. Velmurugan and Srithar [9] studied the factors that influence the performance and the productivity rate of the solar stills, such as temperature differences within the still, inlet water temperature, and angle of the glass cover. As the yield of solar still is low, extensive research has been done toward maximizing the output at a minimal cost [10]. To improve the efficiency of the still, active distillation systems such as solar still coupled with flat plate collectors or parabolic concentrators, still coupled with a hybrid PV system, multistage solar distillation system, and multi-effect solar distillation system have been developed theoretically and experimentally [11]. On studying and comparing various types of solar stills and their different designs [12], a single basin type solar still coupled with an evacuated tube solar collector is chosen to be further analyzed.

The first mathematical model for the heat transfer equations of the evacuated tube collector (ETC) and its analysis was carried in 1976 by Eberlein [13] with air as the working fluid. As shown in Fig. 9.2, water to be purified is supplied to the ETC through the storage. Due to density differences, the heated water inside the tubes rises to the solar still basin. This modified design of the solar still improves the efficiency of the system as the overall heat loss coefficient decreases in the collector.

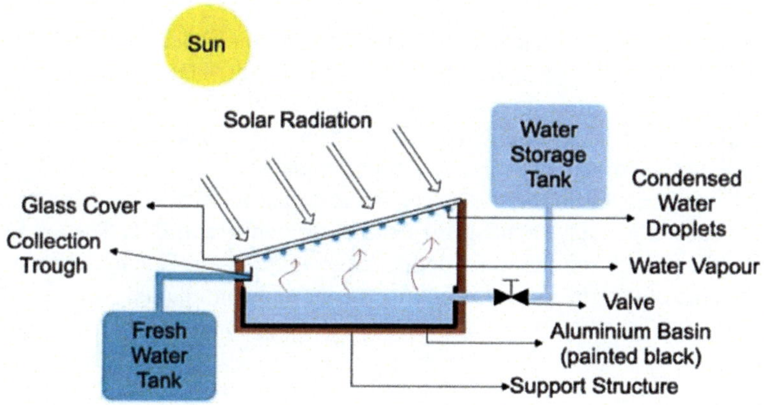

Fig. 9.1 Schematic diagram of a single basin solar still

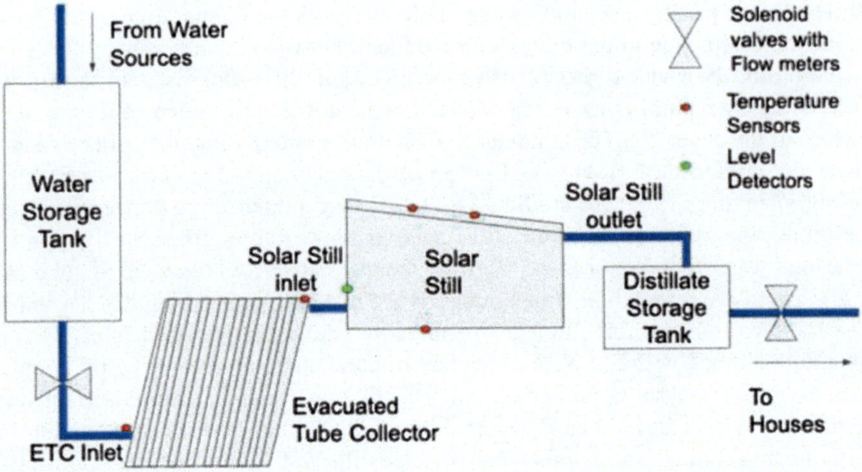

Fig. 9.2 Block diagram of solar still coupled with a U-type evacuated tube collector

9.3 Mathematical Model

The following section discusses the mathematical model of a single slope single basin solar still [5, 14] as the first section, followed by the modeling of a solar still coupled with evacuated tube collector [6] as the second section, to improve the productivity of the system by increasing the temperature difference between the basin water and the glass cover.

The energy balance for each component of single slope basin solar still, as shown in Fig. 9.3, and for ETC system has been considered with the following assumptions: (a) The heat capacity of the insulating materials on the sides and bottom of the basin and the glass cover is negligible, (b) There are no vapor leakages in the solar still unit, (c) Water vapor and dry air are assumed to have a behavior similar to an ideal gas, (d) Water in the basin is static, (e) Physical properties of water and material used are considered to remain unchanged for different temperature, (f) The temperature gradient of the cover and the water depth can be ignored, (g) Condensation of water on the surface of the glass cover is homogeneous and continuous.

9.3.1 Simple Solar Still

Basin Liner. The energy balance equation for the basin is given by

$$a_{\text{lb}} I(t) = q_{\text{cbw}} + q_{\text{tba}} \tag{9.3.1}$$

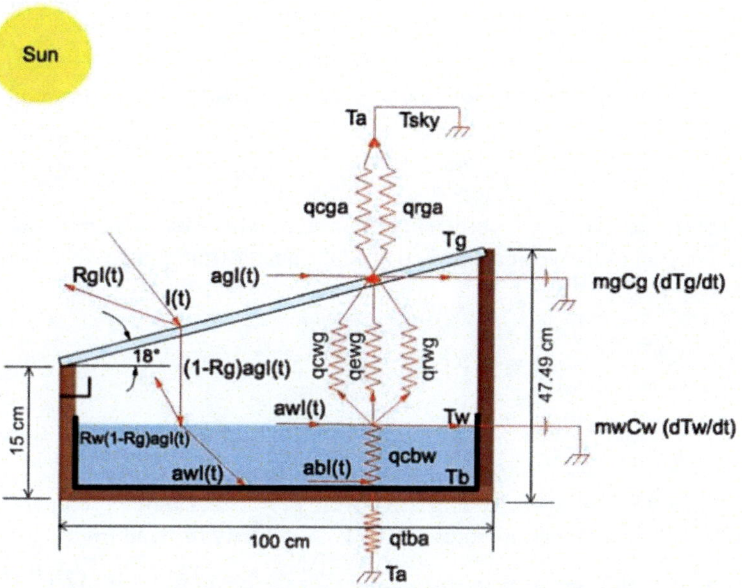

Fig. 9.3 Schematic diagram of the heat transfer analogy of single slope single basin solar still

where a_{1b}—fraction of solar flux absorbed by basin liner; q_{tba}—total heat transfer from basin liner to ambient ($\frac{\mathrm{W}}{\mathrm{m}^2}$).

The internal convective, evaporative, and radiative heat transfer from basin water to glass cover is calculated by

$$q_{\mathrm{cwg}} = 0.884\left[T_{\mathrm{w}} - T_{\mathrm{g}} + \frac{(p_{\mathrm{w}} - p_{\mathrm{g}})(T_{\mathrm{w}} + 273)}{(268.9 \times 10^3 - p_{\mathrm{w}})}\right]^{\frac{1}{3}} (T_{\mathrm{w}} - T_{\mathrm{g}}) \tag{9.3.2}$$

$$q_{\mathrm{ewg}} = 16.276 \times 10^{-3} q_{\mathrm{cwg}}\left(\frac{p_{\mathrm{w}} - p_{\mathrm{g}}}{T_{\mathrm{w}} - T_{\mathrm{g}}}\right) \tag{9.3.3}$$

$$q_{\mathrm{rwg}} = \epsilon_{\mathrm{w}}\, \sigma\left[(T_{\mathrm{w}} + 273)^4 - \left(T_{\mathrm{g}} + 273\right)^4\right] \tag{9.3.4}$$

where T_{w}—temperature of basin water (°C); T_{g}—glass cover temperature (°C); $\epsilon_{\mathrm{w}} = 0.82$—effective emissivity of water surface and glass cover; σ—Stefan–Boltzmann constant W/(m^2 K^4); p_{w}—partial saturated vapor pressure at basin water temperature (N/m^2) and p_{g}—partial saturated vapor pressures at glass cover temperature (N/m^2) given by

$$p_{\mathrm{w}} = \mathrm{e}^{\left(25.317 - \frac{5144}{T_{\mathrm{w}} + 273}\right)} \tag{9.3.5}$$

$$p_g = e^{\left(25.317 - \frac{5144}{T_g + 273}\right)} \quad (9.3.6)$$

The internal convective heat transfer from basin liner to water is calculated by

$$q_{cbw} = h_{cbw}(T_b - T_w) \quad (9.3.7)$$

$h_{cbw} = 250$—heat transfer coefficient by convection from basin liner to water $\left(\frac{W}{m^{2\circ}C}\right)$.

The external radiative and convective heat transfer from glass cover to ambient is calculated by

$$q_{rga} = \in_g \sigma\left[(T_g + 273)^4 - (T_s + 273)^4\right] \quad (9.3.8)$$

$$q_{cga} = (2.8 + 3V_w)(T_g - T_a) \quad (9.3.9)$$

where $\in_g = 0.95$—emissivity of glass cover; $T_a = 35$—ambient temperature (°C), $T_s = 0.0552T_a^{1.5}$—sky temperature (°C); V_w—velocity of wind (m/s).

Basin Water. The energy balance equation for the basin water is given by

$$a_{1w}I(t) + q_{cbw} = m_w c_w f_w + q_{cwg} + q_{ewg} + q_{rwg} \quad (9.3.10)$$

where a_{1w}—fraction of solar flux absorbed by basin water; q_{cbw}—heat transfer through convection from basin liner to water $(\frac{W}{m^2})$; m_w—the mass of water (kg); c_w—specific heat of basin water (J/kg °C); $f_w = \frac{d(T_w)}{dt}$.

Glass Cover. The energy balance equation for the glass cover is given by

$$q_{cwg} + q_{ewg} + q_{rwg} + a_{1g}I(t) = q_{rga} + q_{cga} \quad (9.3.11)$$

where q_{cwg}—heat transfer through convection from basin water to glass cover $(\frac{W}{m^2})$; q_{ewg}—heat transfer through evaporation from basin water to glass cover $(\frac{W}{m^2})$; q_{rwg}—heat transfer through radiation from basin water to glass cover $(\frac{W}{m^2})$; a_{1g}—fraction of solar flux absorbed by glass cover; $I(t)$—solar intensity $(\frac{W}{m^2})$; q_{rga}—heat transfer from glass cover to ambient through radiation $(\frac{W}{m^2})$; q_{cga}—heat transfer from glass cover to ambient through convection $(\frac{W}{m^2})$.

The external total heat transfer from basin liner to ambient is calculated by

$$q_{tba} = h_{tba}(T_b - T_a) \quad (9.3.12)$$

where $h_{tba} = \frac{h_i k_i}{h_i d_i + k_i}$—total heat transfer coefficient from basin liner to ambient $\left(\frac{W}{m^{2\circ}C}\right)$; $h_i = 2.8$—total heat transfer coefficient from the bottom of the basin to ambient $\left(\frac{W}{m^{2\circ}C}\right)$; $k_i = 0.03$—thermal conductivity of insulation $\left(\frac{W}{m^{2\circ}C}\right)$; d_i—thickness of insulation (m).

The fraction of solar radiation absorbed by the glass cover, water, and basin liner is given by

$$a_{1g} = (1 - R_g)a_g \tag{9.3.13}$$

$$a_{1w} = (1 - a_g)(1 - R_g)(1 - R_w)a_w \tag{9.3.14}$$

$$a_{1b} = (1 - a_g)(1 - R_g)(1 - R_w)(1 - a_w)a_b \tag{9.3.15}$$

where $R_g = 0.05$—reflectivity of glass cover; $a_g = 0.05$—absorptivity of the glass cover; $R_w = 0.05$—reflectivity of water; $a_w = 0.05$—absorptivity of water; $a_b = 0.9$—absorptivity of basin liner.

9.3.2 Solar Still Coupled with an Evacuated Tube Collector

Glass Cover. The energy balance equation for the glass cover is given by

$$a_g I_s + h_1(T_w - T_{ci}) = \frac{K_g(T_{ci} - T_{co})}{L_g} \tag{9.3.16}$$

where a_g—the absorptivity of glass cover; I_s—insolation on solar still with the coupling of vacuum tubes $\left(\frac{W}{m^2}\right)$; h_1—total heat transfer coefficient $\left(\frac{W}{m^2}K\right)$; T_{ci},T_{co}—temperatures of inner and outer sides of glass cover (°C); K_g—thermal conductivity of glass cover $\left(\frac{W}{m}K\right)$; L_g—thickness of glass cover (m).

Water Mass. The energy balance equation for the water is given by

$$a_w(1 - a_g)I_s + h_2(T_b - T_w) + Q_u = (MC)_w\frac{d(T_w)}{dt} + h_1(T_w - T_{ci}) \tag{9.3.17}$$

where a_w—the absorptivity of water; h_2—heat transfer coefficient from glass cover to water; T_b—temperature of the basin (°C); $(MC)_w$—the heat stored by the water (kJ/kg); τ—time (s)

$$Q_u = F_R\left[(\alpha\tau)_c I_c - \frac{U_{LC}A_s}{A_{ET}(T_w - T_a)}\right] \tag{9.3.18}$$

where Q_u—heat gained by vacuum tubes on coupling with passive solar still (kJ); F_R—collector heat removal factor; $(\alpha\tau)_c$—the effective transmittance–absorptance product of vacuum tube collector (supposed 0.87 in calculations); I_c—insolation on vacuum tube collector $\left(\frac{W}{m^2}\right)$; U_{LC}—total heat transfer loss coefficient for vacuum collector $\left(\frac{\mathrm{W}}{\mathrm{m}^2}\mathrm{K}\right)$; A_s,A_{ET}—area of solar still and vacuum tube collector (m^2); T_a—ambient temperature (°C).

Basin Liner. The energy balance equation for the basin is given by

$$a_b\left(1-a_g\right)(1-a_w)I = h_2(T_b - T_w) + U_b(T_b - T_a) \tag{9.3.19}$$

where a_b—the absorptivity of the basin; I—insolation on passive solar still $\left(\frac{\mathrm{W}}{\mathrm{m}^2}\right)$; U_b—bottom loss coefficient for passive solar still $\left(\frac{\mathrm{W}}{\mathrm{m}^2}\mathrm{K}\right)$.

Heat transfer rate of convection, evaporation, and radiation, along with the heat transfer coefficients of convection, evaporation and are calculated as follows:

$$q_{cw} = h_{cw}(T_w - T_{ci}) \tag{9.3.20}$$

$$h_{cw} = 0.884\left[(T_w - T_{ci}) + \frac{(p_w - p_{ci})T_w}{(268.9 \times 10^3 - p_w)}\right]^{\frac{1}{3}} \tag{9.3.21}$$

$$q_{ew} = h_{ew}(T_w - T_{ci}) \tag{9.3.22}$$

$$h_{ew} = 16.276 \times 10^{-3} h_{cw}\left(\frac{p_w - p_{ci}}{T_w - T_{ci}}\right) \tag{9.3.23}$$

$$h_1 = h_{rw} + h_{cw} + h_{ew} \tag{9.3.24}$$

$$q_{rw} = \epsilon_{eff}\, F_{12}\sigma\left(T_w^4 - T_{ci}^4\right) \tag{9.3.25}$$

$$h_{rw} = \epsilon_{eff}\, \sigma\left(T_w^2 - T_{ci}^2\right)(T_w + T_{ci} + 546) \tag{9.3.26}$$

where p_w, p_{ci}—water pressure and partial pressure of steam near glass cover; ϵ_{eff}—emissivity of glass cover; F_{12}—form factor; σ—Stefan–Boltzmann constant W/(m^2 K^4).

Hourly distillate output of the still is

$$m = 3600\frac{h_{ew}(T_w - T_{ci})}{L}A_s \tag{9.3.27}$$

The daily distillate output is

$$M_{ew} = \sum_{k=1}^{24} m_k \tag{9.3.28}$$

where k—hours of the day.

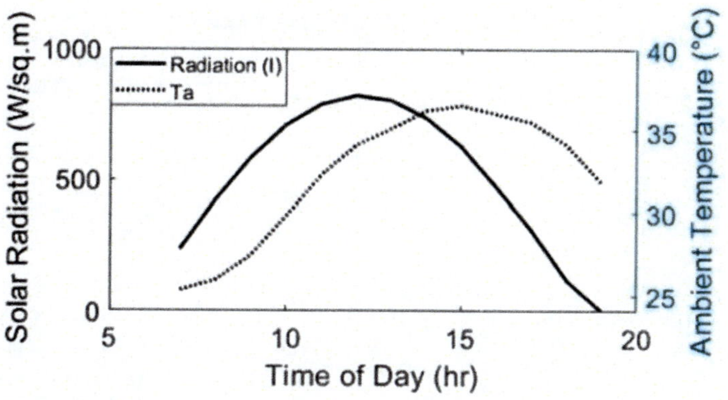

Fig. 9.4 Hourly variation of solar radiation I (W/m^2) and ambient temperature Ta (°C) for May

9.4 Results

The mathematical models discussed in the previous section are simulated in the MATLAB Simulink environment. The performance analysis of a simple solar still and a solar still coupled with ETC is carried out based on varying the parameters affecting its productivity such as solar radiation, thickness of the glass cover, depth of feed water in the basin, area of the basin, and the inlet water temperature.

9.4.1 Solar Radiation

The hourly variation of the ambient temperature and solar radiation at different times of the day during the month of May in Pune district [15] are studied to observe its effect on distillate production as shown in Fig. 9.4. The irradiance cannot be controlled; however, as the productivity of the still is directly proportional to the incident solar radiation, it becomes necessary to position the still such that it receives adequate sunshine [16]. The average radiation for the month of May is given as an input to verify the heat balance Eqs. (9.3.1), (9.3.10), and (9.3.11) in the MATLAB simulation and is used for all the further discussions. It is observed that the radiation is maximum at 12 p.m. with a value of 820.3 W/m^2.

9.4.2 The Thickness of the Glass Cover

Thickness is an important feature that is considered while selecting the glass cover. The transmittance for a 4 mm, 5 mm, and 6 mm thick glass is 0.9, 0.84, and 0.79, respectively [17]. Figures 9.5, 9.6, and 9.7 show that the variation of total daily

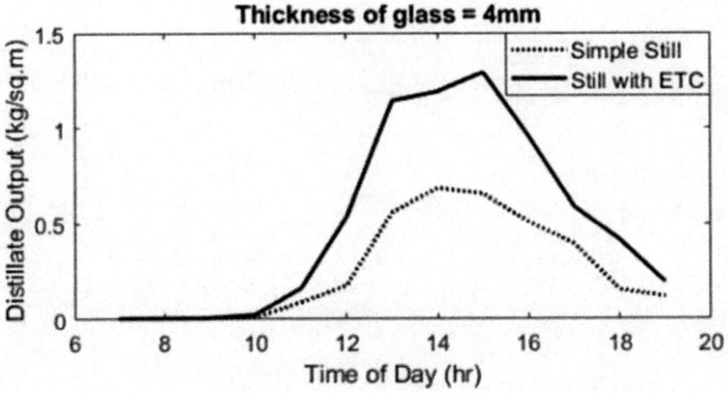

Fig. 9.5 Comparison of hourly variation of the distillate output for glass thickness 4 mm

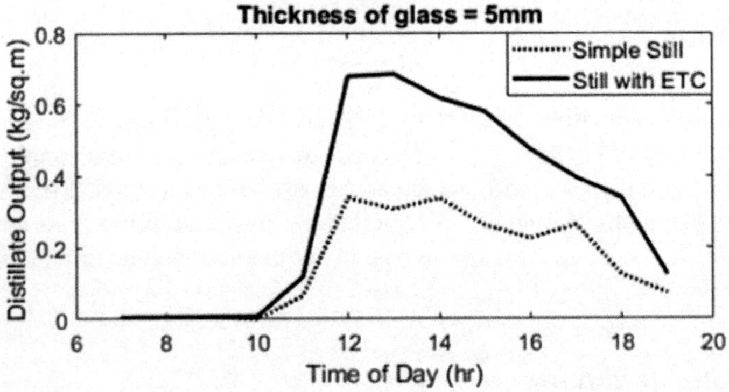

Fig. 9.6 Comparison of hourly variation of the distillate output for glass thickness 5 mm

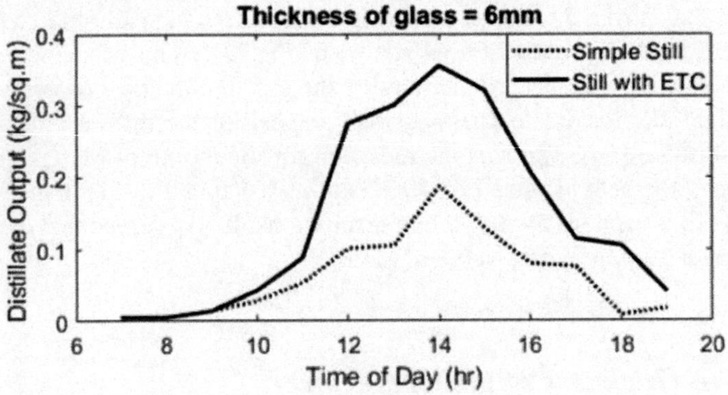

Fig. 9.7 Comparison of hourly variation of the distillate output for glass thickness 6 mm

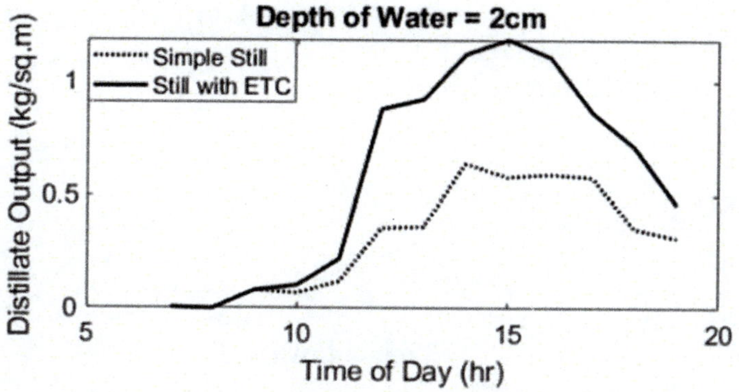

Fig. 9.8 Comparison of hourly variation of the distillate output for 2 cm depth of water

distillate of a simple still is 3.37 kg/m^2 for 4 mm, 2.01 kg/m^2 for 5 mm, and 0.81 kg/m^2 for 6 mm thick glass, respectively, while that of a still coupled with ETC is 6.51 kg/m^2 for 4 mm, 4.03 kg/m^2 for 5 mm, and 1.87 kg/m^2 for 6 mm. An increase in the thickness of glass reduces the efficiency of the still. Since maximum transmittance and minimum absorption and reflection are desired for high productivity, the ideal glass cover thickness is 4 mm [16, 18].

9.4.3 Depth of Water in the Basin

With an increase in the depth of water, the temperature of water decreases due to an increased volume of water, causing a decrease in the rate of evaporation. Thus, the depth of water is another significant factor affecting the production capacity of the still, as the output decreases with an increase in depth of the water [16] as seen in Figs. 9.8, 9.9 and 9.10. The daily distillate output of a simple still is compared with that of a still coupled with ETC for different depths of water. Area of the basin being constant, the only way to decrease the heat absorbed is by decreasing the depth of the water in the basin which also results in higher thermal efficiency [18].

9.4.4 Area of the Basin

The area of the basin is varied to observe its effect on hourly distillate output. The daily output with area of 0.25 m^2, 0.5625 m^2, 1 m^2, 1.5625 m^2, and 3.0625 m^2 for a simple still is 0.84 kg/m^2, 1.89 kg/m^2, 3.37 kg/m^2, 5.26 kg/m^2, and 10.32 kg/m^2, respectively. The output was found to be 1.62, 3.66, 6.51, 10.18, and 19.95 kg/m^2 for a still coupled with ETC. It is seen in Fig. 9.11 that the distillate output increases with the increase in area.

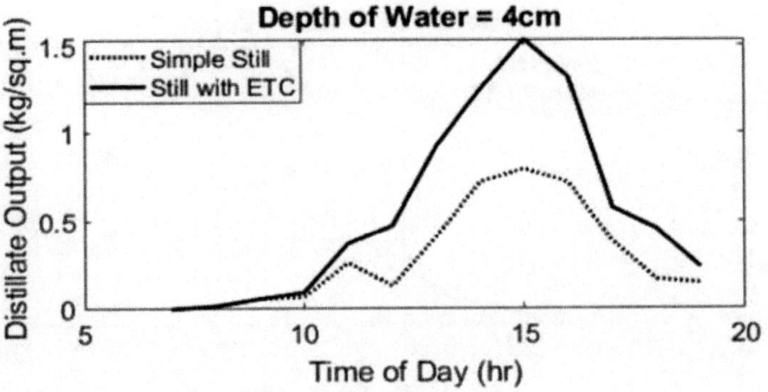

Fig. 9.9 Comparison of hourly variation of the distillate output for 4 cm depth of water

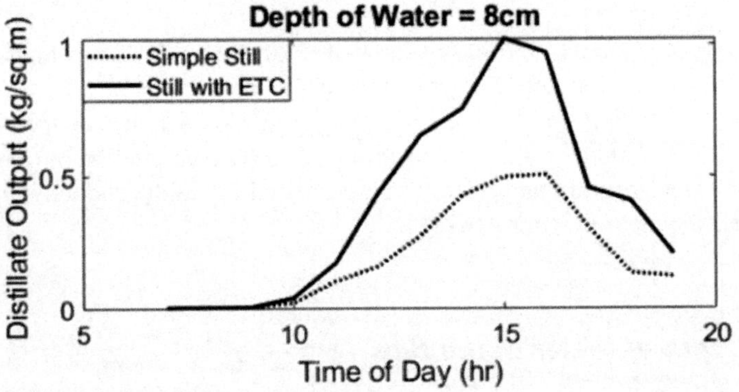

Fig. 9.10 Comparison of hourly variation of the distillate output for 8 cm depth of water

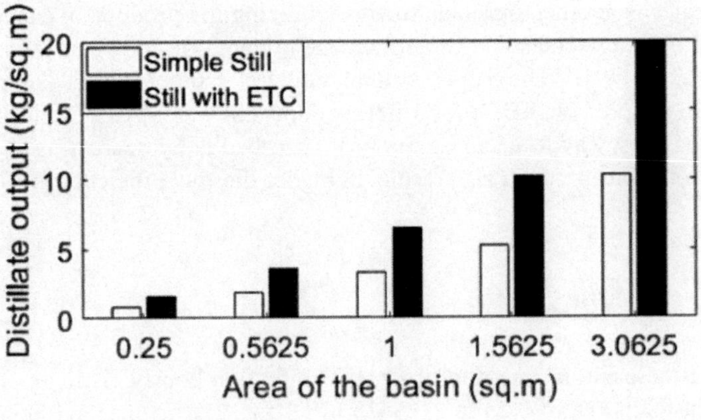

Fig. 9.11 Comparison of the daily distillate output for a simple still and a still coupled with evacuated tubes with a varying basin area

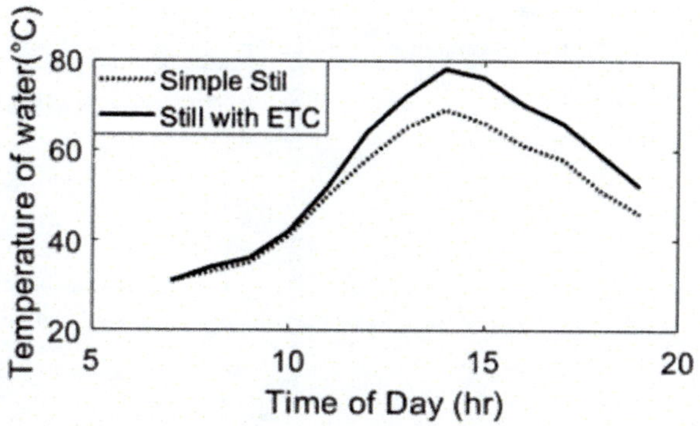

Fig. 9.12 Hourly variation of water temperature (Tw) for simple still and a still coupled with ETC

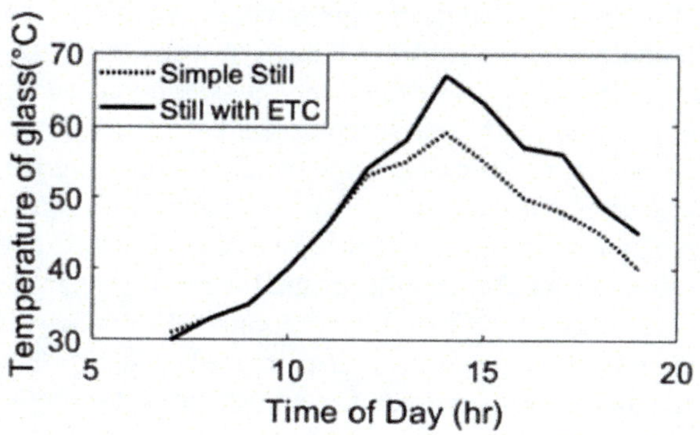

Fig. 9.13 Hourly variation of temperature of glass (Tg) for simple still and still coupled with ETC

9.4.5 Inlet Water Temperature

The increase in temperature of the water in case of a solar still coupled with an evacuated tube collector as compared to that of the simple still is shown in Fig. 9.12. The temperature of glass also increases for a still coupled with ETC as seen in Fig. 9.13. This increases the temperature difference between the glass cover and basin water, consequently increasing the efficiency of the still [10].

The total daily distillate output for a simple still with a glass cover of thickness 4 mm, water depth 6 cm, and a basin area of 1 m^2 is found to be 3.37 kg/m^2 while that of the same still coupled with evacuated tubes is found to be 6.42 kg/m^2 as seen in Fig. 9.14. This clearly shows that the proposed modification of the still increases the output considerably.

9.5 Case Study: Water and Product Survey

Pune is the seventh largest metropolitan city by population and the ninth largest city in the country. Given the fast rate of decline of groundwater level in the last decade,

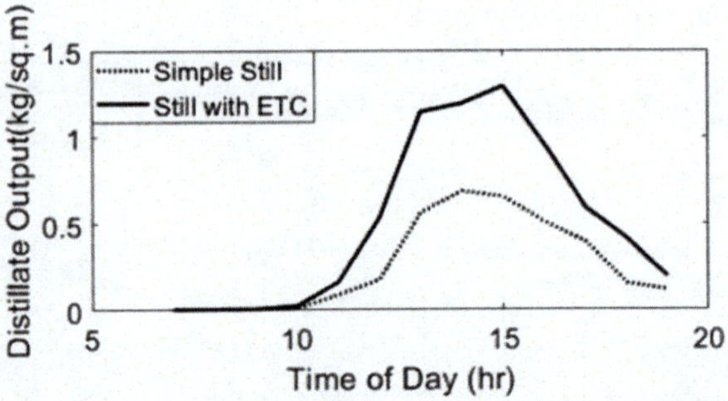

Fig. 9.14 Hourly variation of daily distillate output of a solar still and a still coupled with ETC

many areas in Pune are facing water problems as they are heavily dependent on groundwater for most of their activities. Therefore, a survey was conducted in Pune to understand the current situation of water purification techniques used in various localities such as residential buildings, schools, and hospitals. A questionnaire was prepared to study the different sources of water, the amount of water consumed per person for drinking, bathing, and washing purposes. The survey also intended to learn about the types of commercial purification systems used, along with their maintenance frequencies and costs. According to the results of the survey conducted, 70% of the consumers live in apartments.

Now, a detailed analysis is presented in Table 9.1 of how much each family spends annually on water purification compared to the cost they would have to spend if a solar desalination unit is used. The amount of water produced each day for ten families by commercial water purifiers and single basin solar still was found to be around **94 L** and **60 L**, respectively. A solar still does not lead to any wastage as compared to commercial water purification techniques which waste around 30% of the water

Table 9.1 Cost comparison between individual methods and solar still for water purification

No. of families	Method of water purification	Installation and manufacturing cost per system in Rs	Annual maintenance and operation cost in Rs	Amount of water produced per system in liters	Amount of water wasted in liters	Total cost in Rs
For individual methods of water purification						
7	RO	15,000 * 7	4000 * 7	10 * 7	30 * 7	126,000
2	UV + UF	12,000 * 2	3500 * 2	8 * 2	0	29,000
1	–	0	0	8 * 1	0	0
10	**–**	**129,000**	**35,000**	**94 L/day**	**210 L/day**	**164,000**
For solar still						
10	Solar desalination	**12,250 * 10 = 122,500**	**10 * 10 * 365 = 36,500**	**60 L/day**	**0 L/day**	**159,000**

Table 9.2 Calculation of return of interest for individual methods and solar still

Sr. No.	Various cost parameters	For RO, UV, UF-based water purifiers	For solar still
1.	Fixed cost (instalment and manufacturing)	Rs 129,000	Rs 122,500
2.	Variable cost (maintenance and operation cost)	Rs 35,000/365 = Rs 96 per day	Rs 36,500/365 = Rs 100 per day
3.	Cost of commercially sold drinking water	Rs 20 per liter	Rs 20 per liter
4.	Cost of water produced = Cost of commercial water * Amount of water produced per day	Rs 20 * 94 = Rs 1880 per day	Rs 20 * 60 = Rs 1200 per day
5.	Net profit = Produced water cost − Variable cost	Rs 1784 per day	Rs 1100 per day
6.	Payback period = Fixed cost/Net profit	129,000/1784 = **72 days**	122,500/1100 = **111 days**

supplied as input. Furthermore, solar stills do not consume any electric energy to operate and require minimal cost for its maintenance.

The data obtained in Table 9.1 is used to calculate the return of investment of these purification methods for average-sized ten families as shown in Table 9.2. Thus, solar stills would be beneficial in urban areas and cities where high tariff rates are applied to electricity consumption. The solar stills fulfill the necessary potable water requirements, provide economic relief, and are also energy efficient and environmentally a better option for purification.

9.6 Conclusion

It is estimated that as many as 22 Indian cities will run out of groundwater by 2020. Metropolitan cities especially face water scarcity regularly as it has become difficult to provide clean drinking water in cities with such a growing population. The increased electrical energy consumption in these cities is another major concern that will be addressed by reducing the energy conversion. A desalination-based water purification system will resolve these issues in urban areas as solar energy is a renewable energy source. It can be concluded from the simulations that the solar still coupled with an evacuated tube collector (ETC) produced a higher distillate output as compared to the single basin solar still. This topology can be installed at terraces of buildings in cities to provide potable water to the people. The case study clearly shows that this method of water purification is cost efficient and results in zero wastage of water as opposed to existing water purifiers. The payback period for the solar still is comparatively higher than the membrane-based purification methods;

however, the proposed design will be beneficial in the long run. This prototype requires minimum electrical input for its maintenance and operation and is energy efficient, making it economically feasible for an average-sized family residing in a modern city. Thus, in a developing country like India, which receives annual solar radiation of 200 MW/km^2, this design can be easily accommodated by spending less than what the citizens spend currently on water purification systems. Another aspect is that 68% of India's population resides in rural areas with no supply of electricity and water to fulfill their daily needs. The water purification system proposed in this paper can also be installed in such regions, along with a few modifications to make it more affordable and user friendly.

References

1. R. Saidur, E.T. Elcevvadi, S. Mekhilef, A. Safari, H.A. Mohammed, An overview of different distillation methods for small scale applications. Renew. Sustain. Energy Rev. **15**, 4756–4764 (2011)
2. M. Eltawil, Z. Zhao, L. Yuan, *Renewable Energy Powered Desalination Systems: Technologies and Economics-State of the Art* (IWTC Alexandria, Egypt, 2008), pp. 1099–1136
3. M. Ali, H.E.S. Fath, P.R. Armstrong, A comprehensive techno-economical review of indirect solar desalination. Renew. Sustain. Energy Rev. **15**, 4187–4199 (2011)
4. G.M. Ayoub, L. Malaeb, Developments in solar still desalination systems: a critical review. Crit. Rev. Environ. Sci. Technol. **42**(19), 2078–2112 (2012)
5. A. Agrawal, R.S. Rana, P. Srivastava, Heat transfer coefficients and productivity of a single slope single basin solar still in Indian climatic condition: experimental and theoretical comparison. Resour. Efficient Technol. **4**(3), 466–482 (2017)
6. H.N. Panchal, H. Thakkar, Theoretical and experimental validation of evacuated tubes directly coupled with solar still. Therm. Eng. **63**, 825–831 (2016)
7. H. McCracken, J. Gordes, *Original: Understanding Solar Stills* (VITA), (1985)
8. M. Thimmaraju, D. Sreepada, G.S. Babu, B.K. Dasari, S.K. Velpula, N. Vallepu, *Desalination of Water* (IntechOpen, 2018)
9. V. Velmurugan, K. Srithar, Performance analysis of solar stills based on various factors affecting the productivity—a review. Renew. Sustain. Energy Rev. **15**(2), 1294–1304 (2011)
10. F.M. Ghanim, Mathematical modeling of a solar still. M.S. thesis, Dept. Mech. Eng, University of Khartoum, Sudan (2008)
11. K. Sampathkumar, An experimental study on single basin solar still augmented with evacuated tubes. Therm. Sci. (2012)
12. T. Arunkumar, K. Vinoth Kumar, A. Amimul, R. Jayaprakash, K. Sanjay, Experimental study on various solar still designs. ISRN Renew. Energy (2012)
13. M.B. Eberlein, *Analysis and Performance Predictions of Evacuated Tubular Solar Collectors Using Air as the Working Fluid* (University of Wisconsin, 1976)
14. N. Bao, *The Mathematical Model of Basin-Type Solar Distillation Systems* (IntechOpen, 2019). https://doi.org/10.5772/intechopen.83228
15. Photovoltaic Geographical Information System Homepage, https://re.jrc.ec.europa.eu/pvg_tools/en/tools.html#
16. Md.N.I. Sarkar, A.I. Sifat, S.M. Reza, Md. Sadique, A review of optimum parameter values of a passive solar still and a design for southern Bangladesh. Renew. Wind Water Solar **4**(3) (2012)

17. H. Panchal, Performance investigation on variations of glass cover thickness on solar still: experimental and theoretical analysis. Technol. Econ. Smart Grids Sustain. Energy **1**(7) (2016)
18. A. Santos, E. Hernandez, Experimental evaluation of a single slope solar still. TECCIENCIA **22**(12) (2017)

Chapter 10
Exploring the Role of Lactic Acid Pretreatment in the Saccharification of Guinea Biomass Waste

Meyyappan Geetha Valliammai and Nellaiappan Olaganathan Gopal

Abstract The microbial cellulase is a hopeful technology in meeting out the energy requirement of the people. This study exhibits the efficiency of cellulase in YES5 in converting Guinea biomass waste into high value green compounds. YES5 was molecularly characterized as *Trichosporon asahii*. YES5 exhibited cellulase activity of 21. 12 μmol of glucose min^{-1} on utilizing CMC as carbon source. YES5 was screened for cellulose hydrolyzing capacity via Congo-red agar assay, and the cellulose hydrolyzing capacity was 360%. The cellulose content of Guinea biomass got enhanced to 51.06% on lactic acid pretreatment. Organic acid pretreatment has less environmental impact. The saccharification potential of the isolate *Trichosporon asahii* YES5 in Guinea biomass waste was estimated, and it was maximum on the third day at 33.63%. This study is first of its kind in knowing the positive impact of lactic acid pretreatment in Guinea biomass waste saccharification.

Keywords Microbial cellulase · YES5 · *Trichosporon asahii* · Lactic acid · Saccharification

10.1 Introduction

There is an urge in shifting from non-renewable resources to renewable resources to fulfil the energy requirement. Biopolymers in natural resources are playing a key role in the evolution of value-added products [1]. The driving force towards renewable energy is global warming, greenhouse effect and pollution [2].

Lignocellulosic biomass from forest ecosystem seems to be the best substrate for renewable energy. Cellulose in the waste biomass acts as a suitable substrate for the cellulase. Microbial cellulase is an environmental-friendly system that renders a reasonable yield of free sugars [3]. Cellulases are inducible enzymes secreted by a various microbes on utilizing cellulosic material [4]. The cellulases from different

M. G. Valliammai (✉) · N. O. Gopal
Department of Agricultural Microbiology, Tamil Nadu Agricultural University, Coimbatore, Tamil Nadu 641003, India

R. Seyezhai et al. (eds.), *Recent Trends in Renewable Energy Sources and Power Conversion*, Springer Proceedings in Energy,
https://doi.org/10.1007/978-981-16-0669-4_10

microbes vary in their catalytic potential and stability, which determine the rate of degradation of cellulose [5].

Trichosporon asahii is a species of yeast in the genus *Trichosporon*, a genus of anamorphic fungi in the family Trichosporonaceae. The yeast reported in waste biomass conversion was *Candida tropicalis, Candida intermedia, Pichia guilliermondii, Saccharomyces cerevisiae* and *Trichosporon asahii* [6].

Dilute acid and alkali is one of the most promising biomass pretreatment due to its low cost and retains rest over cellulose in solid residue for the succeeding enzymatic hydrolysis [7]. The pretreatment using organic acid prevents the degradation of substrates and prevents loss of free sugars and inhibitory compounds which further affect the fermentation process [8].

Guinea grass is native to Africa. Approximately 80–100 t/ha of green Guinea fodder is obtained per year, and the high cellulose content in Guinea biomass makes it a promising biomaterial for production of biomass-based green chemicals [9].

In this study, YES5 yeast was isolated from soil, and its efficiency in lactic acid-pretreated Guinea biomass saccharification was noted.

10.2 Materials and Methods

10.2.1 Isolation, Screening and Molecular Identification of Cellulolytic Yeast YES5

YES5 was isolated from Eastern Ghats region (N 10. 29 363°) (E 77. 70 674°) of Thandikudi, Tamil Nadu, India, via serial dilution and pour plating technique. One gram of the soil sample was suspended in 9 mL of water and used for isolating the cellulolytic yeast using the carboxymethyl cellulose (CMC) agar medium (1% of peptone, 1% of CMC, 0.1% of K_2HPO_4, 0.03% of $MgSO_4$, 0.02% of $(NH_4)_2SO_4$, 0.02% of NaCl, 2% of agar, 6.2–6.8: pH supplemented with chloramphenicol (100 μg/mL)) and incubated at 28 °C for three days. YES5 was screened for cellulase activity by using the Congo-red agar medium [10]. YES5 was phylogenetically characterized by 18S rRNA technique. Nearly full length of the 18S rRNA gene was amplified from the genomic DNA using the forward primer NL-1 (5′-GCA TAT CAA TAA GCG GAG GAA AAG-3′) and the reverse primer NL-4 (5′-GGT CCG TGT TTC AAA GAC GG-3′) [11].

10.2.2 Cellulase Assay

YES5 was grown in CMC medium at 28 °C for three days, and the culture supernatant was collected by centrifugation at 12,000 rpm for 8 min. The culture supernatant (150μL) was mixed with 200 μL of 1% CMC solution and incubated at 50 °C for

30 min. Then, 3 mL of DNS reagent was added and incubated at 100 °C for 5 min. Absorbance was determined at 575 nm against the reagent blank. One unit of enzyme activity was defined as the quantity of enzyme that released 1 μmol of reducing sugar per minute per millilitre [12].

10.2.3 Lactic Acid Pretreatment on Guinea Biomass Waste

Five grams of the 50 μm sieved Guinea biomass were taken into a series of 250 ml conical flask, 100 ml of 3.0% concentration of lactic acid was added to each flask to hydrolyze the substrates, and the flasks were kept in autoclave at 121 °C for 30 min followed by sudden depressurization by fully opening the steam exhaust valve of autoclave. The compositional analysis of Guinea biomass before and after pretreatment was determined [13].

10.2.4 YES5 in Lactic Acid-Pretreated Guinea Biomass Waste Saccharification

One gram of the dried Guinea biomass powder was added to 100 mL of CMC broth in which Guinea biomass was replaced for CMC as a source of carbon. YES5 was inoculated, and the saccharification reaction was allowed at 30 °C and 120 rpm for five days. The culture supernatant was isolated for every 24 h to monitor the release of reducing sugar pattern. The experiments were performed in triplicate, and the saccharification yield was noted [14].

10.3 Results and Discussion

10.3.1 YES5 Colony Characteristic and Phylogenetic Analysis

The yeast isolate YES5 was creamy and glossy colony. YES5 was identified belonging to the species *Trichosporon asahii* based on 18S rRNA sequencing. YES5 got its NCBI accession number as MK640632. Cellulose hydrolyzing capacity of YES5 was qualitatively observed in the form of halo zone formation and quantitatively measured as 360% (Table 10.1). Yeast such as *Candida tropicalis* isolated from soil sample was reported as biomass utilizing cellulase producer [15].

Table 10.1 Cellulose hydrolyzing efficiency of YES5

Yeast isolate	NCBI accession number	Cellulose hydrolyzing capacity (%)
YES5	MK640632	360

Table 10.2 Impact of lactic acid pretreatment on Guinea biomass waste

Biomass waste	Before pretreatment cellulose content	After pretreatment cellulose content
Guinea biomass waste	46.64 ± 0.18	51.06 ± 1.02

10.3.2 Cellulase Activity of YES5

Cellulase activity of YES5 was quantitatively measured as 21.12 μmol of glucose min^{-1}. The most important component in biomass is cellulose and is converted into value-added products by the enzyme cellulase [16]. Biomass-acting enzymes are promising due to their suitability for industrial applications [17].

10.3.3 Effect of Lactic Acid Pretreatment on Guinea Biomass Waste

The cellulose content of Guinea biomass before and after lactic acid pretreatment was quantified as 46.64% and 51.06%, respectively (Table 10.2). This shows the positivity of lactic acid pretreatment in the enhancement of cellulose content. Among the organic acid pretreatment on rice straw, 4% lactic acid treatment was effective [8].

10.3.4 Saccharification of Guinea Biomass Waste

Saccharification of Guinea biomass was performed using *Trichosporon asahii* isolate YES5, and the saccharification percentage was evaluated for every 24 h. The maximum saccharification percentage was quantified as 33.63%, and it was observed on day 3 (Fig. 10.1). *Pichia stipitis* has been used for the saccharification of sugarcane bagasse [18]. *Saccharomyces cerevisiae* has been used for the saccharification of cotton residues [19].

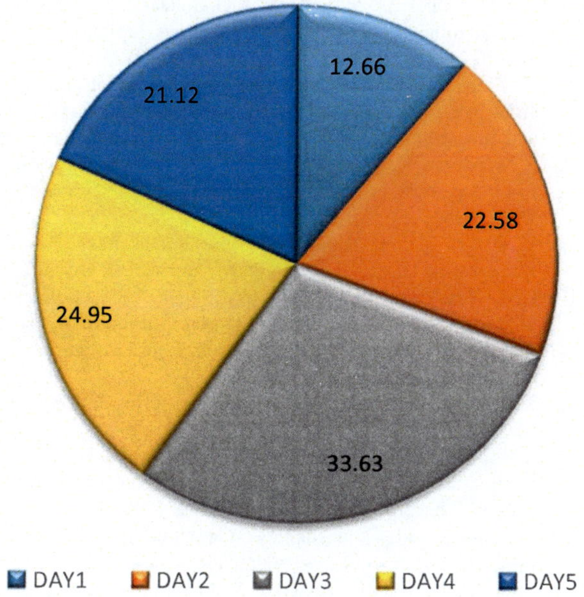

Fig. 10.1 Saccharification efficiency of YES5 on lactic acid-pretreated Guinea biomass

10.4 Conclusion

A cellulolytic isolate, YES5, isolated from soil sample, was found to belong to *Trichosporon asahii*. 3% lactic acid pretreatment was effective in Guinea biomass utilization. The yeast isolate YES5 proved to be an effective strain that could perform saccharification of Guinea biomass waste. Thus, this study adds the significance of organic acid pretreatment in biomass utilization, which can be further utilized in green chemistry.

References

1. P. Vanhala, I. Bergström, T. Haaspuro, P. Kortelainen, M. Holmberg, M. Forsius, Boreal forests can have a remarkable role in reducing greenhouse gas emissions locally: land use-related and anthropogenic greenhouse gas emissions and sinks at the municipal level. Sci. Total Environ. **557**, 51–57 (2016)
2. R. Kumar, C.E. Wyman, Does change in accessibility with conversion depend on both the substrate and pretreatment technology? Bioresour. Technol. **100**(18), 4193–4202 (2009)
3. A.K. Chandel, G. Chandrasekhar, M.B. Silva, S. Silvério da Silva, The realm of cellulases in biorefinery development. Crit. Rev. Biotechnol. **32**(3), 187–202 (2012)
4. R.K. Sukumaran, R.R. Singhania, A. Pandey, Microbial cellulases-production, applications and challenges. J. Sci. Ind. Res. **64**, 834–844 (2005)
5. R. Sindhu, M. Kuttiraja, P. Binod, K.S. Rajeev, P. Ashok, Bioethanol production from dilute acid pretreated Indian bamboo variety (*Dendrocalamus* sp.) by separate hydrolysis and fermentation. Ind. Crops Prod. **52**, 169–176 (2014)

6. R.F. Schwan, E.G. Almeida, M.A.G. Souza-Dias, L. Jespersen, Yeast diversity in rice–cassava fermentations produced by the indigenous Tapirapé people of Brazil. FEMS Yeast Res. **7**(6), 966–972 (2007)
7. K. Karimi, G. Emtiazi, M.J. Taherzadeh, Ethanol production from dilute-acid pretreated rice straw by simultaneous saccharification and fermentation with *Mucor indicus, Rhizopus oryzae*, and *Saccharomyces cerevisiae*. Enzyme Microb. Technol. **40**(1), 138–144 (2006)
8. F. Momayez, K. Karimi, S. Karimi, I.S. Horváth, Efficient hydrolysis and ethanol production from rice straw by pretreatment with organic acids and effluent of biogas plant. RSC Adv. **7**(80), 50537–50545 (2017)
9. Y. Wong, Z. Lam, Bioethanol from Guinea grass (*Panicum maximum*) and ragi tempeh as fermentation starter using bioreactor and shaker. Orient. J. Chem. **35**(4), 1302–1312 (2019)
10. C.W. Hendricks, J.D. Doyle, B. Hugley, A new solid medium for enumerating cellulose-utilizing bacteria in soil. Appl. Environ. Microbiol. **61**(5), 2016–2019 (1995)
11. S. Harju, H. Fedosyuk, K.R. Peterson, Rapid isolation of yeast genomic DNA: Bust n' Grab. BMC Biotechnol. **4**(1), 1–6 (2004)
12. T.M. Wood, K.M. Bhat, Methods for measuring cellulase activities, in *Methods in Enzymology*, vol. 160 (Elsevier, 1988), pp. 87–112
13. J. Ntumba, A. Mulula, K. Kashishi, M. Mifundu, R. Robiette, K. Taba, Physicochemical properties of diacetylenic light fuel oil from congolese oleaginous plant Ongokea gore (Hua) Pierre. J. Appl. Chem. **2017** (2017)
14. C. Uma, C. Muthulakshmi, D. Gomathi, V. Gopalakrishnan, Fungal invertase as aid for production of ethanol from sugarcane bagasse. Res. J. Microbiol. **5**(10), 980–985 (2010)
15. M.G. Valliammai, N.O. Gopal, R. Anandham, Mining the prospective of *Candida tropicalis* YES3 in Napier biomass saccharification. Biomass Conv. Bioref. 1–11 (2020)
16. J. Lu, X. Li, R. Yang, L. Yang, J. Zhao, Y. Liu, Y. Qu, Fed-batch semi-simultaneous saccharification and fermentation of reed pretreated with liquid hot water for bio-ethanol production using *Saccharomyces cerevisiae*. Bioresour. Technol. **144**, 539–547 (2013)
17. C.J. Duan,J.X. Feng, Mining metagenomes for novel cellulase genes. Biotechnol. Lett. **32**, 1765–1775 (2010)
18. M.W. Lau, C. Gunawan, V. Balan, B.E. Dale, Comparing the fermentation performance of *Escherichia coli* KO11, *Saccharomyces cerevisiae* 424A (LNH-ST) and *Zymomonas mobilis* AX101 for cellulosic ethanol production. Biotechnol. Biofuels **3**(1), 11 (2010)
19. H.G. Yücel, Z. Aksu, Ethanol fermentation characteristics of *Pichia stipitis* yeast from sugar beet pulp hydrolysate: use of new detoxification methods. Fuel **158**, 793–799 (2015)

Chapter 11
Optimization Techniques of Two-Phase Level Converter Ally with Photovoltaic System and Particle Swarm Scheme for Renewable Energy Systems

R. Felshiya Rajakumari and Mrunal Deshpande

Abstract Special techniques are implemented on DC-DC Converter by renewable energy applications. Due to goal for renewable energy sources, the proposed work investigates the double stage system method for renewable applications. The lead role of using coupled inductor in two-segment scheme is: it will shrink the dimension of the structure, produce high system response, reduce ripple performance work, and it will provide high efficiency of the solar PV system. It consists of a no. of components which are alike in boost converter and they are connected in parallel for the proposed topology. The two-phase structure has the similar segments and switching frequency. The proposed work deals with the various parameters to find out and the simulation is done in MATLAB/Simulink. Further the two-phase structure will conglomerate with solar system applications and the MPPT technique such as particle swarm optimization, perturb and observe method is introduced in this paper. The proposed converter, IBC, PV model, P&O, PSO, is simulated in MATLAB/Simulink software.

Keywords Interleaved boost converter · Renewable energy · Ripple · Efficiency · PV panel · Perturb and observe method · Particle swarm optimization technique

11.1 Introduction

The increased environmental pollution leads to the enlargement of traditional energy assets such as PV, wind which is the high turn over of converters. There are so many benefits of using energy resources in the research work. Consequently, using these resources in the converters, it overcomes many of the environmental problems. In this

R. Felshiya Rajakumari (✉)
Department of Electrical and Electronics Engineering, Karpagam Institute of Technology, Coimbatore, India

M. Deshpande
Department of Electrical and Electronics Engineering, SSN College of Engineering, Chennai, India
e-mail: mrunald@ssn.edu.in

R. Seyezhai et al. (eds.), *Recent Trends in Renewable Energy Sources and Power Conversion*, Springer Proceedings in Energy,
https://doi.org/10.1007/978-981-16-0669-4_11

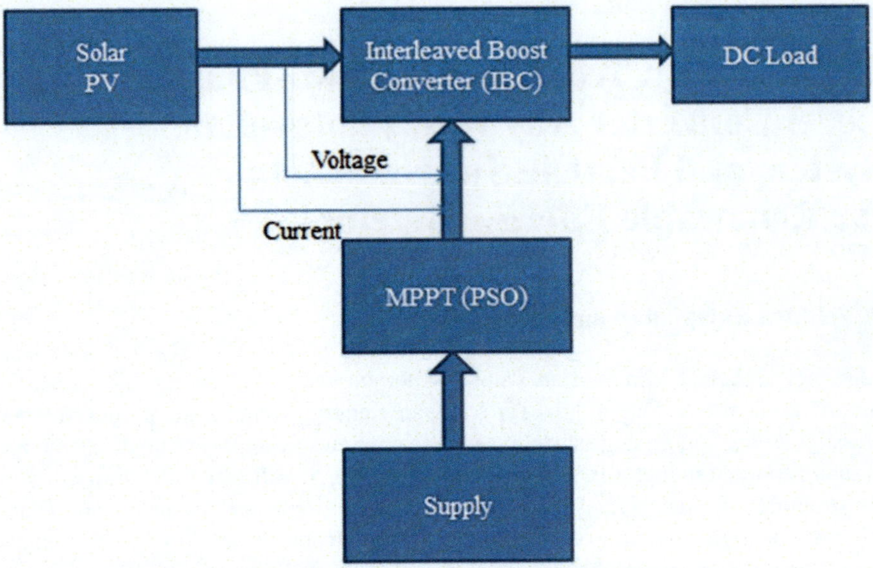

Fig. 11.1 Indicates the pictorial representation of two-phase structure

research exertion, the two-phase scheme conglomerates with PV and MPPT method [1, 2]. There will be so many drawbacks of using other converters because it will produce low voltage gain and low efficiency, but in this work two-phase model is implemented and it results with a high voltage gain and high efficiency. The goal of two-phase structure shrinks the ripple techniques is shown in Fig. 11.1.

Section 11.1 treats with the Introduction of work, Sect. 11.2 discuss with the interleaved boost converter, Sect. 11.3 discusses about the PV modeling. Section 11.4 presents with the MPPT method of particle swarm optimization. Section 11.5 deals simulation outcomes of the converters, and Sect. 11.6 presents the computation of two-phase scheme factors. Section 11.7 initiates the conclusion of the whole work.

11.2 Interleaved Boost Converter

The graphic plan of two-stage scheme is shown in Fig. 11.2. It is designed with the components of two inductor, two diode, two switches, and these are connected in parallel with one another. Two plugs are provided with the gate signal and it is out of phase 180°. When the regulator 1 gets turn ON, the inductor current IL1 gets increases linearly, but when the regulator 2 action was absent, diode D1 gets ON, the energy will get stored in inductor. The same operation will take place in other half of the cycle [3]. The pulse generator is removed by a segmentation with a proper alteration of $\frac{360}{m}$, consider $m = 2$. The two-phase structure drives in four types.

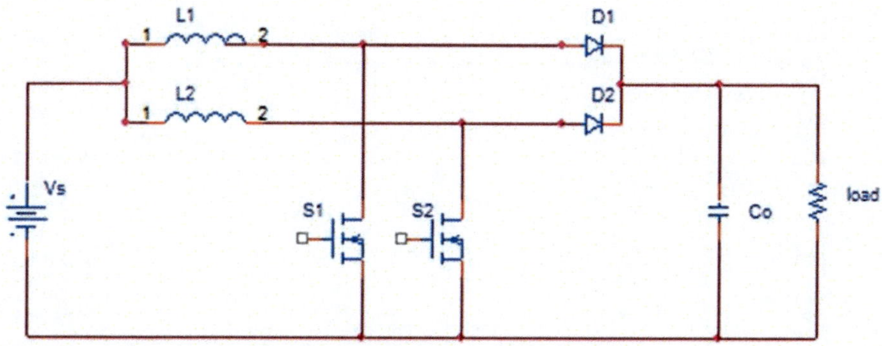

Fig. 11.2 Graphic plan of two-segment structures

11.2.1 Methods of Operation for Interleaved Boost Converter

Method 1

During method I operation, both the regulator operation will take place and both the diodes will be in OFF position, the schematic diagram of method I operation of two-phase scheme is exposed in Fig. 11.3.

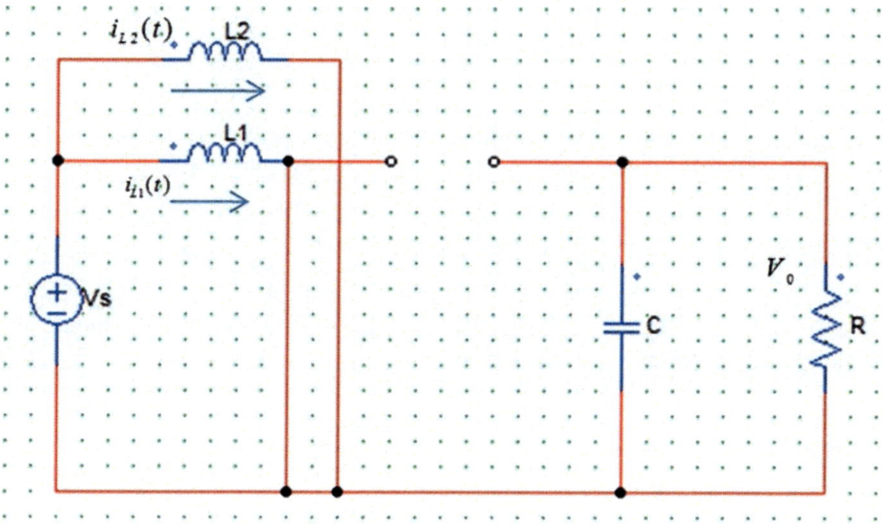

Fig. 11.3 Method I—schematic circuit of two-phase scheme

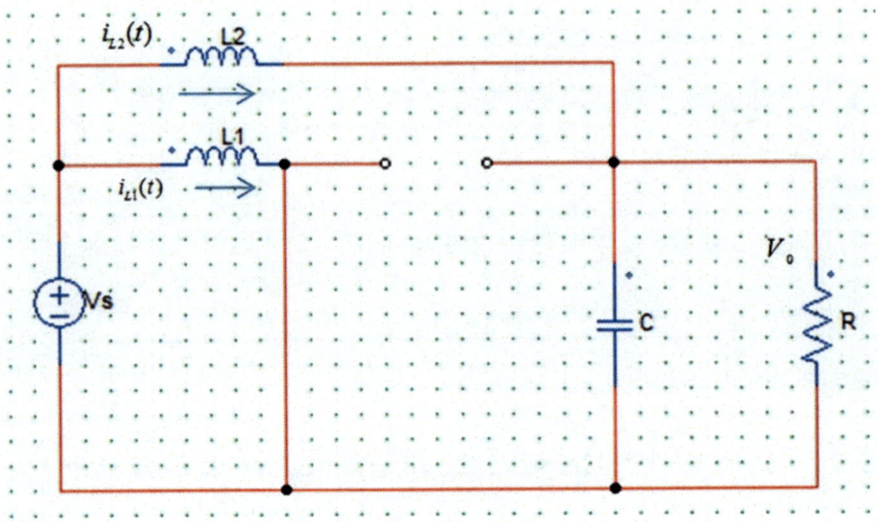

Fig. 11.4 Method 2—schematic circuit of two-phase scheme

Method 2

In method 2 operation, regulator 1 operation takes place, correspondingly the complementary diode acts. The operation will hold for next cycle, i.e., vice versa. The schematic diagram of method 2 operation of two-phase scheme is exposed in Fig. 11.4.

Method 3

During method 3 conditions, the input side of the regulator 1 will be in OFF spot, the regulator 2 operation proceeds. Similarly diode D2 action acts with the help of switch S2. Figure 11.5 demonstrates the schematic circuit of method 3 operation of two-phase scheme.

Method 4

Figure 11.6 displays the schematic circuit of method 4 operation of two-phase scheme. In this condition, both the regulators action will be held OFF; eventually the diode operation will take place in method 4.

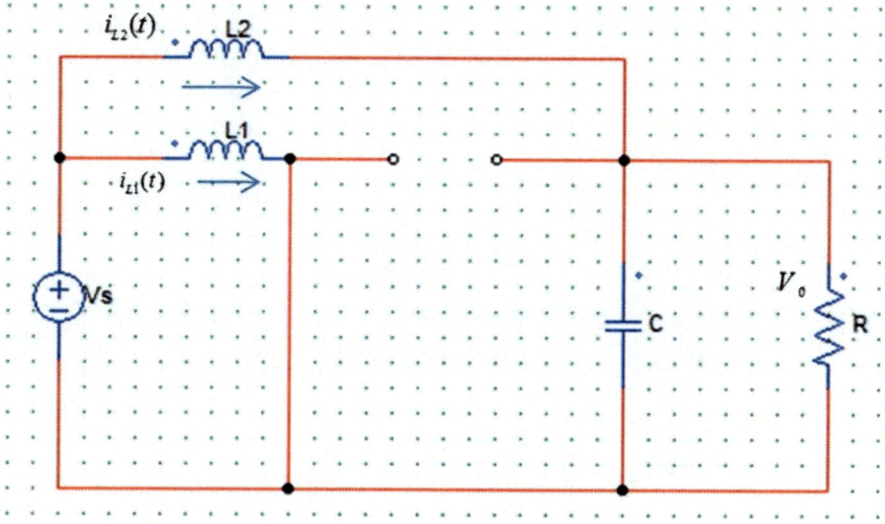

Fig. 11.5 Method 3—schematic circuit of two-phase scheme

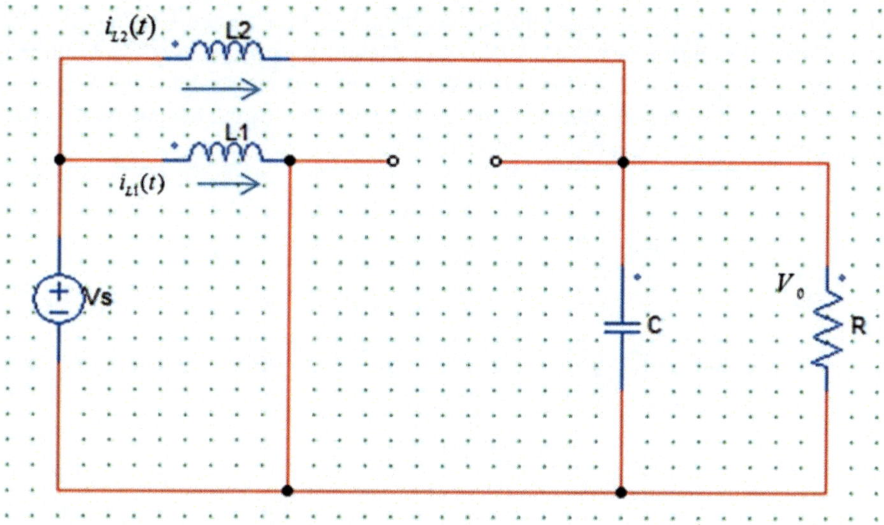

Fig. 11.6 Method 4—schematic circuit of two-phase scheme

11.2.2 Calculations for Two-Phase Method

Output voltage for IBC:

$$V_0 = \frac{V_S}{1 - \alpha} \tag{11.1}$$

Inductance formula for IBC:

$$L = \frac{V_S(V_0 - V_S)}{f_S \Delta I V_0} \tag{11.2}$$

Capacitor formula:

$$C = \frac{I_0 \alpha}{f \Delta V_c} \tag{11.3}$$

Where, L = Inductor, C = Capacitor, f = Switching Frequency, V_s = Source Voltage, V_0 = Load Voltage, α = Duty Cycle of the Converter.

11.3 PV Modeling

Figure 11.7 show the equivalent circuit of PV model. Recently photovoltaic power generation gaining popular for the applications and the solar system used to transform input energy into electric energy [4, 5]. The solar cell components are coupled in sequence and collateral to get the perfect output voltage and current [7]. The PV modeling is simulated in MATLAB/Simulink. Figure 11.8 displays the Simulink diagram of PV panel. Finally power versus voltage, current versus voltage characteristics of solar panel is simulated and it is exposed in Figs. 11.9 and 11.10. The constraints value for PV model is listed in Table 11.1.

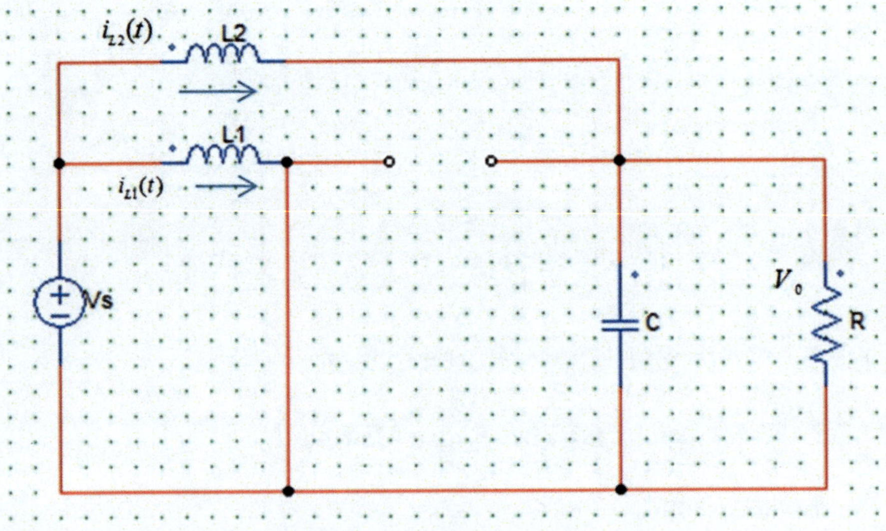

Fig. 11.7 Equivalent circuit of PV cell

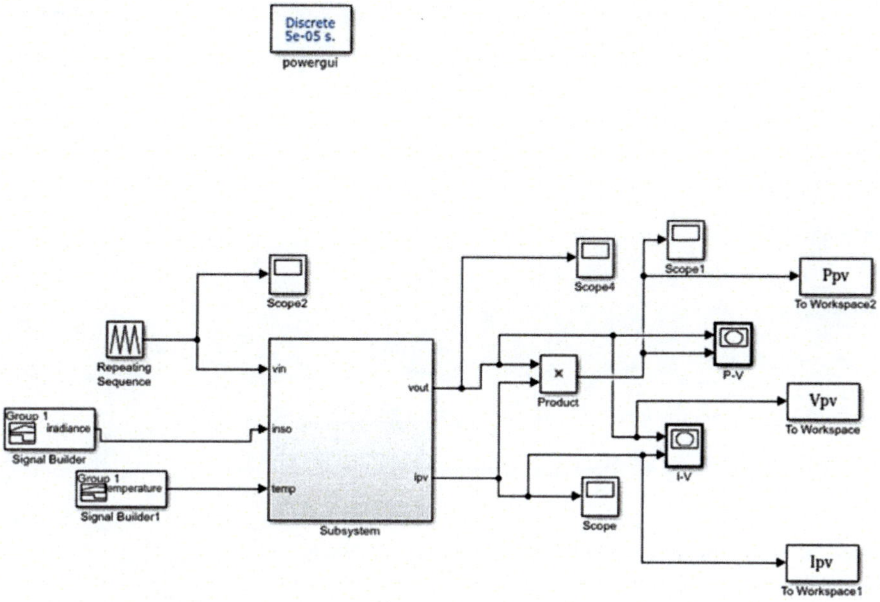

Fig. 11.8 Simulink plan of PV panel

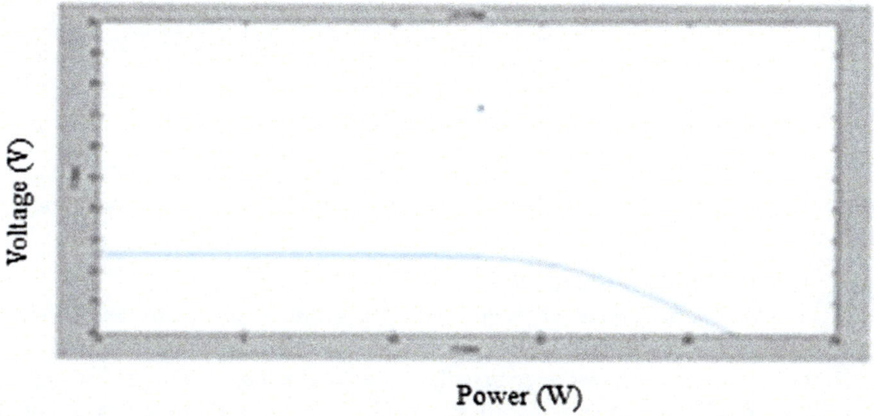

Fig. 11.9 Power versus voltage characteristics of PV model

Figure 11.9 indicates the P–V characteristics of PV model. From the plot, it is inferred that the power is constant with respect to voltage and at a particular period it suddenly decreases.

Figure 11.10 indicates the V–I characteristics of the PV model. From the plot, it is inferred the voltage rises linearly with respect to current and at a particular point it suddenly drops to an instant.

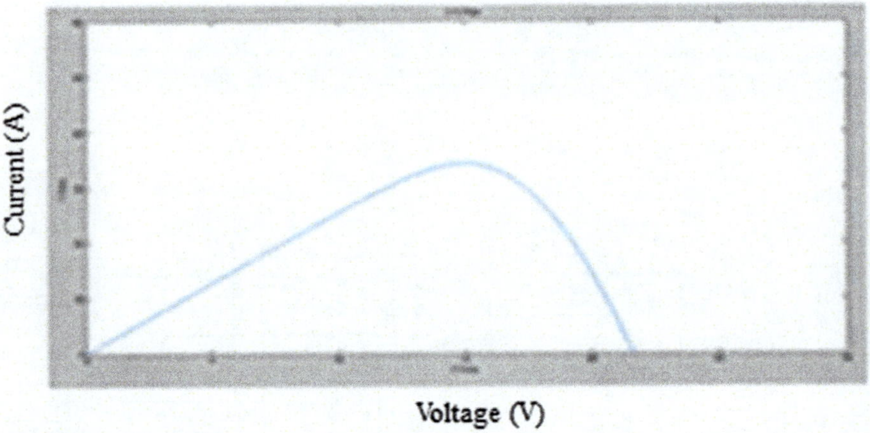

Fig. 11.10 Voltage versus current characteristics of PV model

Table 11.1 Constraints for PV model design

Constraints	Standards
Rated power	10 W
Voltage at maximum power (V)	16.8 V
Current at maximum power (I)	1.19 A
Open circuit voltage (V)	21.4 V
Short circuit current (I)	0.9 A

11.4 Maximum Power Point Tracking

In this broadside work, the MPPT techniques particle swarm optimization is discussed.

(A) **Particle Swarm Optimization**

Particle swarm optimization is used as a spinning procedure for DC-DC converter. The PSO regulator uses a populace of constituent part to achieve the necessary optimization to gain the output at the source voltage. It is implemented with the number of particles, and it is higher-level method to optimize a function. Therefore, PSO method is used to improve the system solution. One of the advantages of the particle swarm optimization is it is used to reduce the number of particles. Compared with other MPPT method, PSO is very simple for performing calculation process and it will produce high tracing efficiency for interleaved boost converter. The flowchart and detailed explanation for particle swarm optimization are discussed in reference section [4–8].

Algorithm for Particle Swarm Optimization

The power $P_{pv,i}$ is premeditated by enlarging the restrained voltage and current. Then the process proceeds to check whether this duty cycle value will result in a better value by evaluating the following equation:

$$P_{pv,i} > P_{pv,i-1} \tag{11.4}$$

If (4) is satisfied, then the value ($p_{best,i}$) is updated; otherwise, $p_{best,i}$ retains its contemporary value. $P_{pv,i}$ is then checked against the power of the other particles to see if the global value (gbest).

(B) **Perturb and Observe Method**
In perturb and observe method, a slight agitation is familiarized in this scheme. This agitation causes the power of the solar module fluctuations. If the power increases due to the agitation, then the perturbation is continued in that direction. After the peak power is reached, the power at the next instant shrinkages. When the steady state is reached, the method fluctuates around the peak point. In order to keep the power deviation small, the agitation size is kept very small. The flowchart and detailed explanation for perturb and observe method are discussed in reference section [9].

Algorithm for Perturb and Observe Method

The scheme is constantly agitated by the deviation in operational point which is positive or else the direction of agitated is swapped. if the change power and voltage are changed in positive or negative values, then the duty cycle is reduced in order to produce the consecutive cycle of agitation. Similarly, if voltage is positive and the power is negative or vice versa then the duty cycle is increased for the next rotation of perturbation. The gains of perturb and observe schemes are simple organization, easy implementation, and vital parameters.

11.5 Simulation Outcomes of Two-Phase Scheme

Finally, the two-phase model is combined with PV and MPPT, and it is simulated in software process and the wave shape is displayed. From the simulation wave shape, various parameters such as ripple process were found out. The simulation parameters for the interleaved boost combine with PV are listed in Table 11.2 (Fig. 11.11).

Figure 11.12 indicates the load voltage for two-phase model converter combined with solar and PSO MPPT, the load voltage displays 24 V.

Figure 11.13 performs the output or load voltage ripple (OVR) for two-phase scheme combine with solar and PSO MPPT, the OVR result is 0.34 V.

Figure 11.14 provides the input current for two-phase scheme combined with solar and PSO MPPT; result for input current is shown as 0.27 A.

Table 11.2 Simulated values for IBC combine with solar

Parameters	Values
Input voltage	12 (V)
Inductor	50 (μH)
Output capacitor	720 (μF)
Load	100 (Ω)

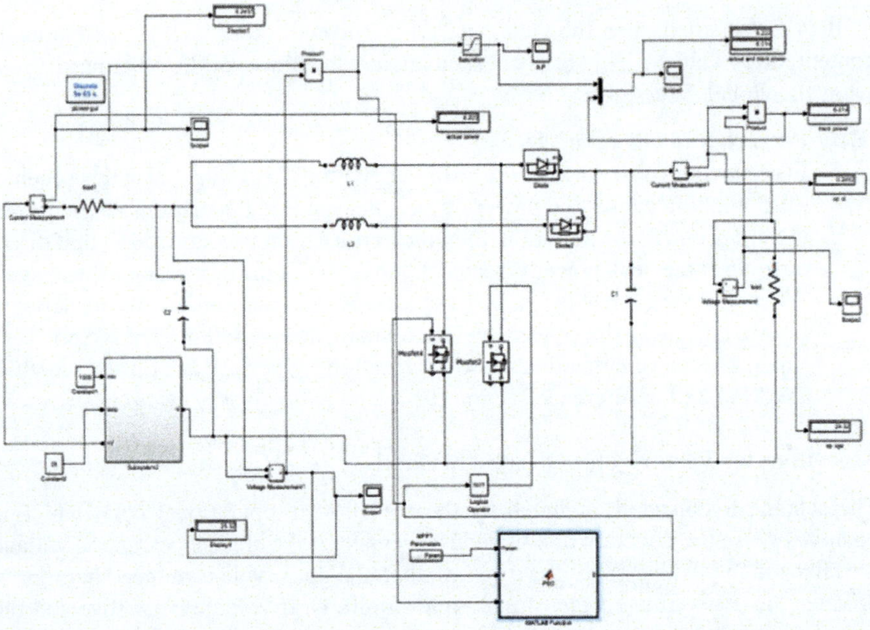

Fig. 11.11 Simulink picture of two-phase model combines with solar and PSO MPPT

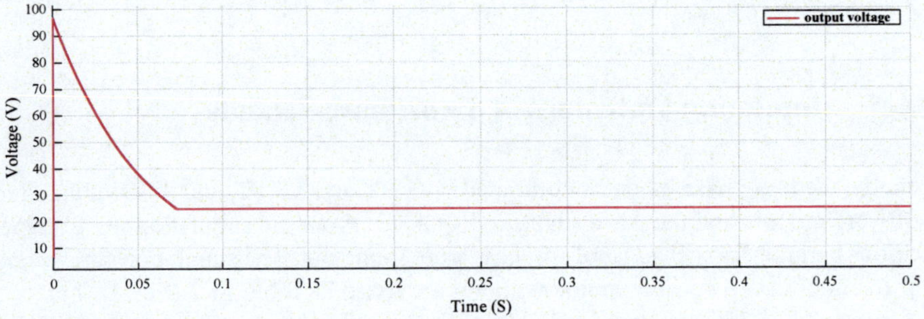

Fig. 11.12 Load voltage of two-phase model combines with solar and PSO MPPT

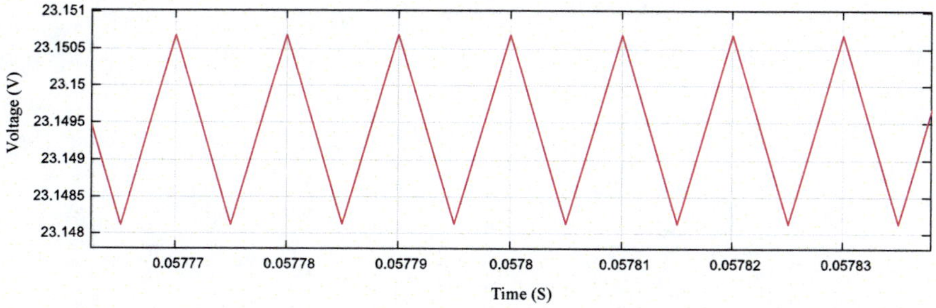

Fig. 11.13 Output voltage ripple of interleaved boost converter combined with PV and PSO

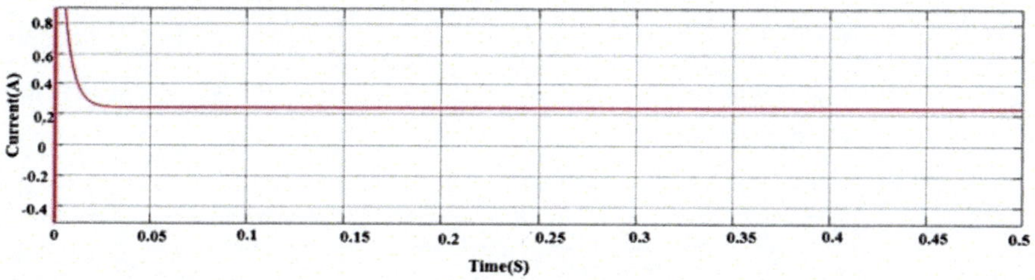

Fig. 11.14 Input current of two-phase model combined with solar and PSO MPPT

Figure 11.15 indicated the input current ripple (ICR) for two-phase order combined with solar and PSO MPPT, the ICR produced as 0.04 A.

Figure 11.16 displays the inductor current ripple (ICR) for two-phase scheme combined with solar and PSO MPPT; the ICR produced as 0.45 A.

Figure 11.17 displays the actual power versus track power for two-phase scheme combined with solar and PSO MPPT; the track power produce 6.214 W and the actual power result 6.8 W (Fig. 11.18).

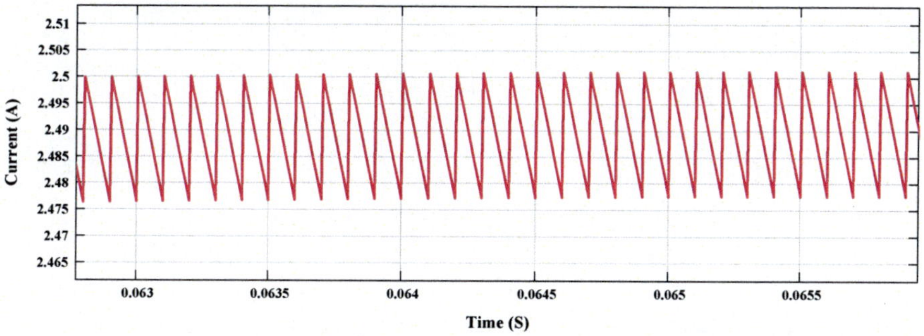

Fig. 11.15 Input current ripple of two-phase order combined with solar and PSO MPPT

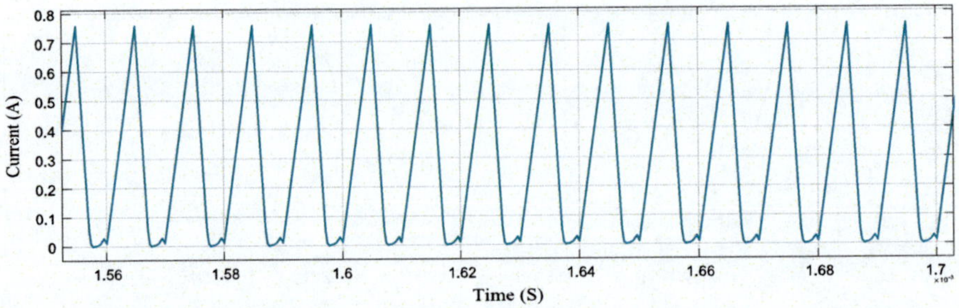

Fig. 11.16 Inductor current ripple of two-phase model combined with solar and PSO MPPT

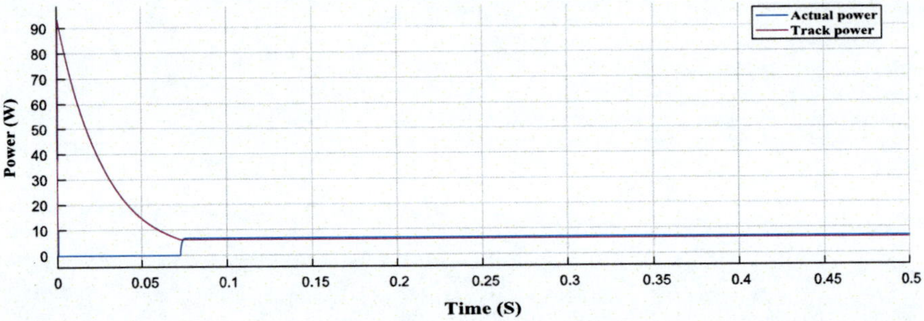

Fig. 11.17 Actual power versus track power of two-phase model combine with solar and PSO MPPT

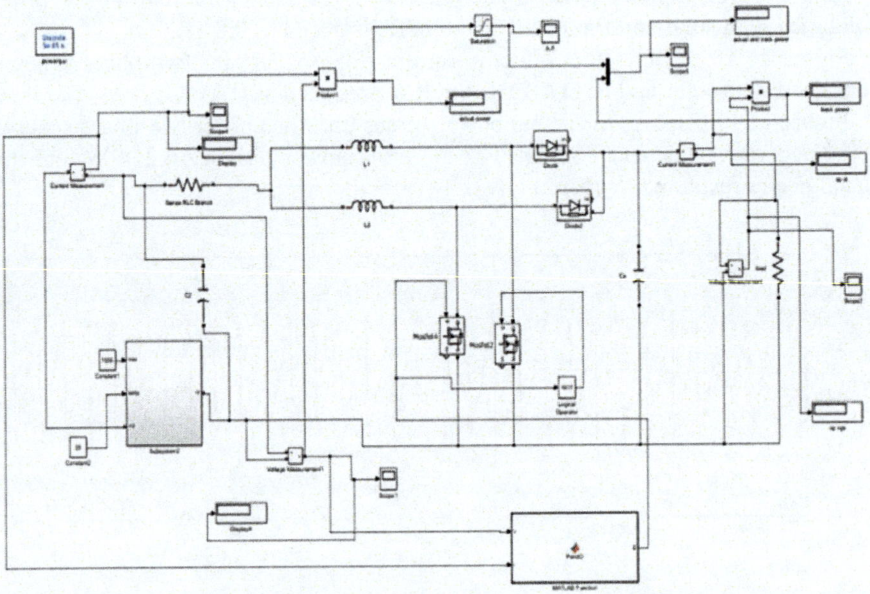

Fig. 11.18 Simulink picture of two-phase model combine with solar and P&O MPPT

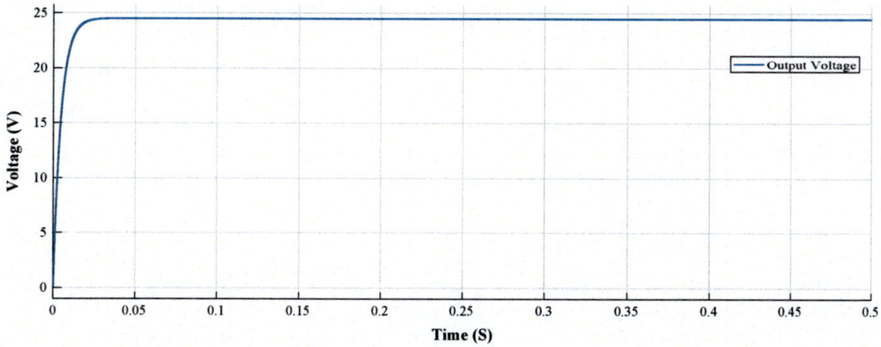

Fig. 11.19 Load voltage of two-phase model combine with solar and P&O MPPT

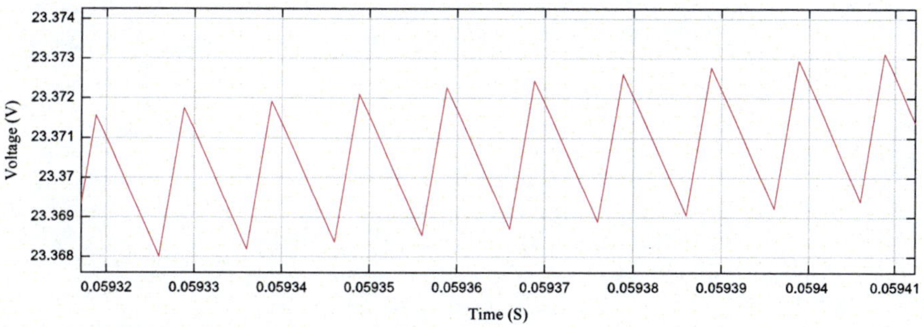

Fig. 11.20 Output voltage ripple of interleaved boost converter combined with solar and P&O MPPT

Figure 11.19 indicates the load voltage for two-phase model converter combined with solar and P&O MPPT, the load voltage displays 24 V.

Figure 11.20 performs the output or load voltage ripple (OVR) for two-phase scheme combined with solar and P&O MPPT, the OVR result is 0.4 V.

Figure 11.21 displays the inductor current ripple (ICR) for two-phase scheme combined with solar and P&O MPPT; the ICR produced as 0.85 A.

Figure 11.22 displays the actual power versus track power for two-phase scheme combined with solar and P&O MPPT; the track power produce 6.214 W and the actual power result 6.8 W.

Figure 11.23 indicated the input current ripple (ICR) for two-phase order combined with solar and P&O MPPT, the ICR produced as 0.34 A.

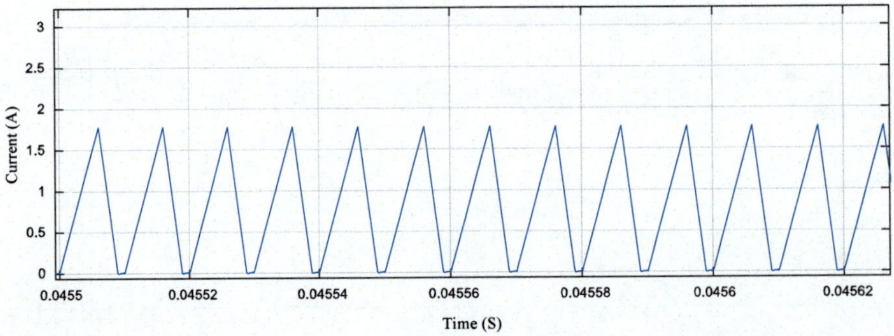

Fig. 11.21 Inductor current ripple of two-phase model combined with solar and P&O MPPT

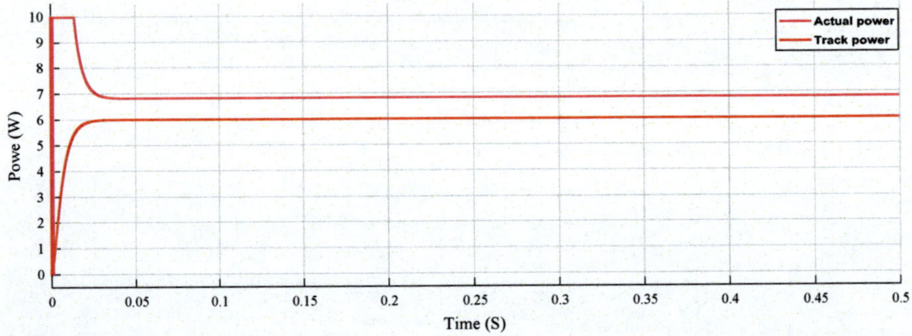

Fig. 11.22 Actual power versus track power of two-phase model combined with solar and P&O MPPT

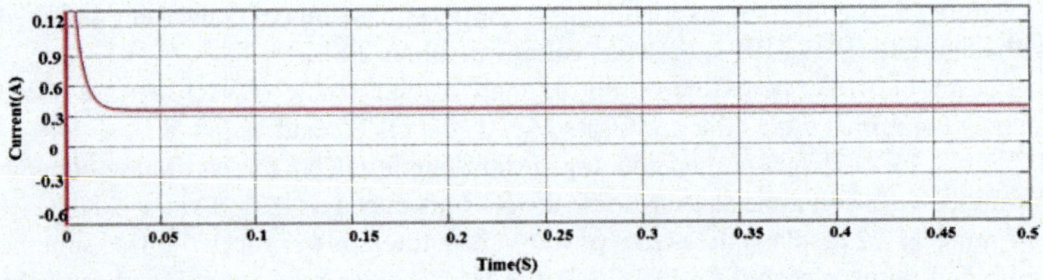

Fig. 11.23 Input current ripple of two-phase order combined with solar and P&O MPPT

11.6 Computation of Two-Phase Pattern Factors Combined with Solar and MPPT

The performance evaluation factors of two-phase system combined with solar and MPPT and these assessments are listed in Table 11.3.

Table 11.3 Performance factors of two-phase scheme combine with solar and MPPT

Parameters	IBC interface with PV and PSO	IBC interface with PV and P&O
Output voltage ripple (V)	0.34	0.4
Input current ripple (A)	0.04	0.34
Inductor current ripple (A)	0.45	0.85
Actual power (W)	6.6	6.8
Track power (W)	5.9	6.2
Tracking efficiency (%)	89	85

From the tabulation, the two-phase model combines with solar and PSO MPPT produces the high tracking efficiency as 89%.

11.7 Conclusion

Two-segment system suites high benefits for renewable energy applications. This work implements the interleaved boost converter interface with PV and maximum power point tracking method such as PSO and P&O. Simulation takes place in MATLAB/Simulink model. The performance parameters such as output voltage ripple, input current ripple, inductor current ripple, and track power versus actual power were performed. From the obtained results, we can conclude that, the ripple technique provides low values in interleaved boost converter and the tracking efficiency results high in two-segment arrangement.

References

1. B.M. Alharbi, M.A. Alhomim, et al., An efficient high voltage gain using two-stage cascaded interleaved boost converter for solar PV system with MPPT technique, in *2020 IEEE Power and Energy Society Innovative Smart Grid Technologies Conference*, 7 May 2020
2. M. Chorishiya, D.K. Palwalia, Analysis of interleaved coupled inductor ultra-boost converter for solar PV applications, in *2019 2nd International Conference on Power Energy, Environment and Intelligent Control*, 3 Feb 2020
3. K. Latha Shenoy, C. Gurudas Nayak, et al., Design and implementation of interleaved boost converter. Int. J. Eng. Technol. **9**. ISSN: 2319-8613 (2019)
4. T. Pavithra, Design and analysis of boost converter based PSO method for MPPT extraction in photovoltaic system. Int. J. Mod. Trends Sci. Technol. **3**(5) (2017). ISSN: 2455-3778
5. D.J. Baby, P. Baburaj, Solar PV fed interleaved boost sepic converter. Int. J. Innov. Res. Electr. Electron. Instrum. Control Eng. **5**(1) (2017). ISSN: 2321-2004

6. J. Divya Navamani, K. Vijayakumar, et al., Study on high step-up DC-DC converter with high gain cell for PV applications, in *7th International Conference on Advances in Computing & Communication, ICACC-2017*, 22–24 Aug 2017
7. N. Selvaraju, et al., Two-phase interleaved boost converter using coupled inductor for fuel cell applications, in *2017 International Conference on Alternative Energy in Developing Countries and Emerging Economics*, 25–26 May 2017
8. V. Pradeepakumara, N. Patil, Renewable energy based interleaved boost converter. Int. J. Innov. Sci. Eng. Technol. **3**(8) (2016). ISSN: 2348-7968
9. M. Patil, A. Deshpande, Design and simulation of perturb and observe maximum power point tracking using MATLAB/Simulink, in *2015 International Conference on Industrial Instrumentation and Control*, 9 July 2015
10. P. Chitra, R. Seyezhai, Basic design and review of two phase and three phase interleaved boost converter for renewable energy system. Int. J. Appl. Sci. (2014). ISSN: 2394-9988
11. N. Coruh, S. Urgun, et al., A simple and efficient implementation of interleaved boost converter, in *6th International Conference on Industrial Electronics and Applications* (2011)
12. Y.-C. Hsieh, T.-C. Hsueh, et al., An interleaved boost converter with zero-voltage transition. IEEE Trans. Power Electron. **24**(4) (2009)

Chapter 12
IoT-Based Home Automation Using Raspberry Pi

Leo Raju, G. Sowmya, S. Srividhya, S. Surabhi, M. K. Retika, and M. Reshmika Janani

Abstract The idea of sustainable energy imposed people to think of a new alternative way to conserve energy. With the development of technology and Internet, a new sensational advancement known as Internet of Things (IoT) was created. This paper mainly focuses on energy conservation through automation for which sensors and a microprocessor are used. The appliances are sensed and controlled using Raspberry Pi for optimal use of electricity in the building. The unnecessary use of electricity due to carelessness of the customer is identified and can be controlled remotely using the mobile application.

Keywords Raspberry Pi · Home automation · Internet of Things (IoT) · Energy management

12.1 Introduction

Internet of Things (IoT) is emerging as one of the most important concepts for making the idea of "smart city" a reality [1, 2]. It finds applications in various sectors including agriculture, retail, health, automobile, etc. The most popular application of IoT is found in smart grids and home automation [3, 4].

Home automation aims to monitor and control the lighting, temperature, humidity and motion of that particular place by means of sensors [5]. All the sensors are ideally connected to a microprocessor, thus making the home automation system, which in turn are connected to a central hub or cloud, and the entire system is controlled by the user by means of an app [6, 7]. Here, Ubidots is used as a cloud platform which is used to receive data from Raspberry Pi and stores it [8, 9]. The proposed system is not limited to home alone. It can be used wherever automation is applicable like offices, labs and banks [10].

This paper presents an advanced method of home automation by using Raspberry Pi as the microprocessor, several sensors and an app which enables the user to

L. Raju (✉) · G. Sowmya · S. Srividhya · S. Surabhi · M. K. Retika · M. Reshmika Janani
SSN College of Engineering, Kalavakkam, Chennai 603110, India
e-mail: leor@ssn.edu.in

R. Seyezhai et al. (eds.), *Recent Trends in Renewable Energy Sources and Power Conversion*, Springer Proceedings in Energy,
https://doi.org/10.1007/978-981-16-0669-4_12

control the electrical appliances, thereby controlling energy consumed in their house. The paper contains information about the proposed system, hardware and software requirements and implementation.

12.2 Proposed System and Block Diagram

In this system, Raspberry Pi 3B+ is used as a connecting bridge between the sensors and the cloud. The Internet connection is established for automation and cloud computation using IoT as shown in Fig. 12.1.

This figure represents the working block diagram of this project. Different sensors—LDR sensors, PIR sensors, current sensors, temperature and humidity sensors—are connected at various points in the house. These sensors require very little current, and the usage of these sensors can reduce energy consumption drastically. The information from the sensors is received by Raspberry Pi which is then sent to the cloud. Based on the information, the device can be controlled by the user with the help of an Android applications. This information is sent back to the Pi from cloud which, in turn, serves as the output to the actuators or relays, controls the devices as per the user's convenience and hence optimizes energy consumption in house. The components are given in Table 12.1.

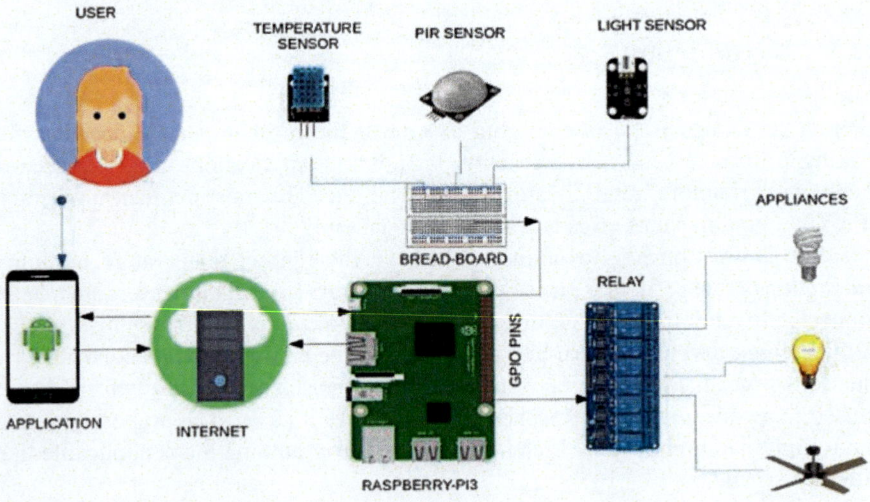

Fig. 12.1 Working block diagram of the project

Table 12.1 Hardware requirements

S. No.	Components	Necessity
1	Raspberry Pi 3 B+	Microprocessor
2	LDR	To detect change in light intensity
3	PIR	To detect infrared radiations
4	Current sensor	Detects current flowing through the circuit
5	Temperature—humidity sensor	Detects change in temperature and moisture content in the atmosphere
6	Relay module	Actuator

Fig. 12.2 Hardware model

12.2.1 Hardware Implementation

The hardware model is assembled, and connections are given as shown in Fig. 12.2.

In our hardware model, we have connected Raspberry Pi to PIR sensor, LED module and motor driver. This is a prototype for the larger scale home automation. We have controlled the sensors and actuator (motor driver) using Raspberry Pi. Now, the data obtained from this will be sent to the cloud platform so as to control from anywhere.

12.2.2 Hardware Prototype

Figure 12.3 shows a simple prototype of a working model of our proposed idea. Here, we used an LED, DC motor and L293D motor driver (actuator) which is connected to Raspberry Pi to control it. The motor driver and LED are connected to GPIO pins of Raspberry Pi. We code in Python IDLE. From time library, we import the sleep function to set proper delay for simultaneous control. We send signals HIGH or LOW from microprocessor to the driver. The driver is in turn connected to the

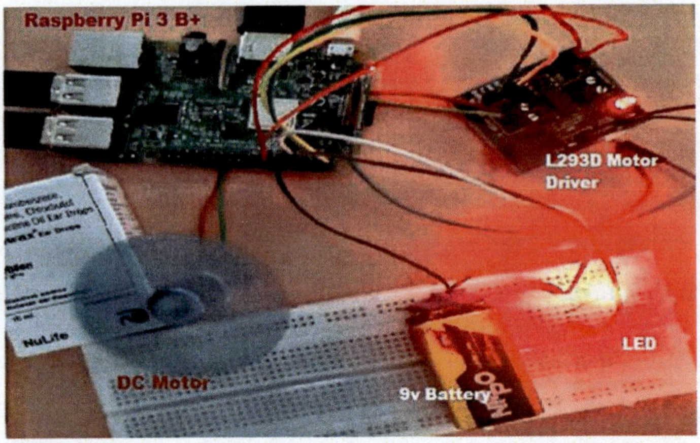

Fig. 12.3 Simple prototype

motor and actuates the motor in FORWARD mode when signal is HIGH and LED is also simultaneously turned on. When signal is LOW, motor is stopped, and LED is turned off. This simple prototype is a miniature version of the working hardware model of the paper. The displayed output can also be sent as an alert message in the mobile application during future enhancements to make it convenient for the user. The prototype and the programming outputs are shown in Figs. 12.3 and 12.4, respectively.

The prototype is shown in Fig. 12.3. Here, the motor driver is used to regulate voltage and send control signals to the DC motor. The 9 V battery is used to power the motor driver. The LED is connected to Raspberry Pi. When Pi sends actuating signal, the LED and the DC motor are turned on, and we get output in display as shown in Fig. 12.4. In this figure, the various control options of the motor and the LED actions are shown.

Fig. 12.4 Prototype programming output display

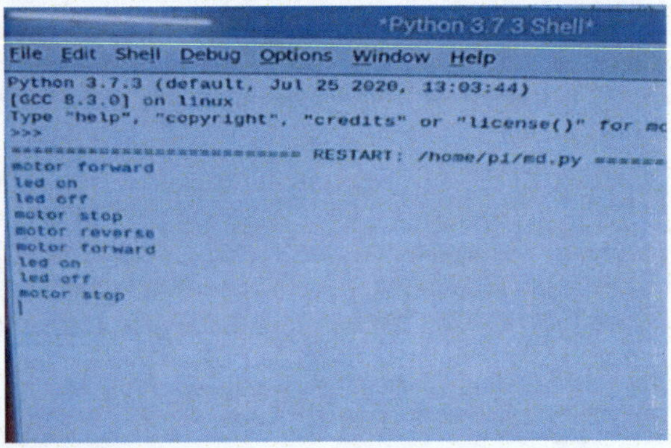

12.3 Software Implementation

12.3.1 *Internet of Things (IoT)*

IoT is the capability to sense, respond and work automatically whenever exposed to a change from the environment it is kept in, without the need of humans. Internet of Things (IoT) has simplified our life and has been incorporated in various industries in various fields. Any thing can be sensed, analysed and controlled remotely across the world. In an IoT system, sensors will send sensed data to a control centre which will decide according to this data and sends command to actuators as a response. In our project also, various sensors send the sensed input to the Raspberry Pi which sends them to the cloud and hence to the app by which user can control the output of the electrical appliances.

12.3.2 *Cloud Computing—Ubidots*

Using cloud, we can send and access data remotely since the information in not stored locally in any device but is stored in data server from which data can be accessed from anywhere. In the system, we use Ubidots Cloud platform. Ubidots is a cloud platform which allows us to send data from a device to cloud by the use of Internet. In Ubidots dashboard, several variables of devices can be added, and sensed values from sensors can be stored in the variables. Values from the cloud can also be accessed by user in app by connecting the app with Ubidots platform. In the Ubidots dashboard, three variables are created and the values sensed by the sensors are stored in these variables.as shown in Fig. 12.5.

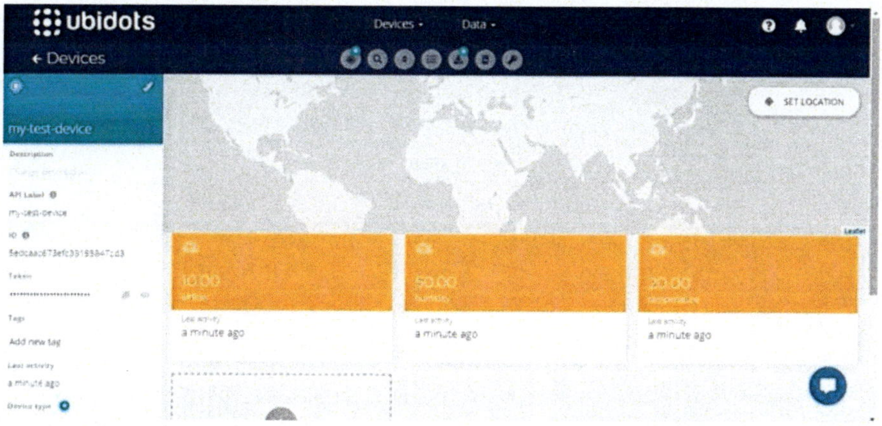

Fig. 12.5 Ubidots dashboard

12.3.3 Smart Mobile Application

The Raspberry Pi is connected to sensors which read the real-time data from the environment and send it to Pi which in turn sends the data to cloud. This data stored in the cloud platform is accessed through our app. Thus, the app obtains the current status of the house, i.e., all the devices which are currently using electricity and also the power consumed by each, and shows that to the user. Then, the user can control which appliances to on or off with the app based on information provided. Thus, we provide an app-based user control. Hence, we can optimize energy consumption in house and regulate the electricity bill cost of the house.

12.4 Machine Learning—A Intelligent Data Analysis

It helps in predicting values without human intervention. We used multiple regression and implemented in Scikit larn. Using decision tree regression algorithm, a graph is plotted as shown in Fig. 12.6. This algorithm gives an accuracy of 99%.

In this figure, two input features are considered for load prediction, namely time and unit price at every hour. The points represent load and its dependency on time and price. From this, we can find how much energy is consumed and also predict the future energy usage. The prior knowledge of energy consumption leads to better planning for optimal usage and also shifts towards a more sustainable way of managing energy. We can also include several other features to improve the accuracy of prediction.

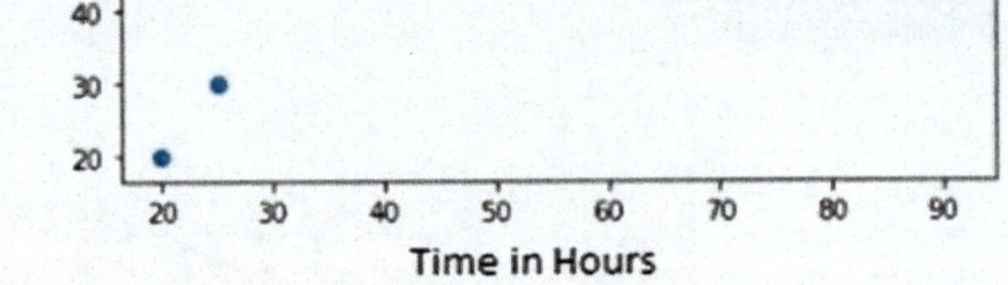

Fig. 12.6 Machine learning data analysis

12.5 Conclusion

We have presented an IoT-based home automation system which will work effectively to monitor the energy consumed in our household and will conserve energy. This system is wholesome, renewable and proficient. The system not only reduces energy consumption benefiting economically, but also conserves the energy which could be used for the future, hence benefitting socially and thereby leading to sustainable growth of our earth with the available resources. We can further extend our project to find out the areas which have high level of electricity usage during certain hours of a day on daily basis and propose a system to conserve and reduce the electricity usage during those maximal usage hours and ensure optimal use of the resources at various levels and institution.

References

1. A.-R. Al-Ali, Java based home automation system. IEEE Trans. Consum. Electron. **50**(2), 498–504 (2004)
2. V. Patchava, H.B. Kandala, P.R. Babu, A smart home automation technique with Raspberry Pi using IoT, in *International Conference on Smart Sensors and Systems (IC-SSS), Bangalore*, pp. 1–4 (2015)
3. J. Chhbra, P. Gupta, IoT based smart home design using power and security management system, in *1st International Conference on Innovation and Challenges in Cyber Security (ICICCS)* (2016)
4. C. Mahapatra, A.K. Moharana, L.C.M. Victor, Energy management in smart cities based on Internet of Things: peak demand reduction and energy savings. Sensors **17**(12):2812 (2017)
5. K. Chooruang, K. Meekul, Design of an IoT energy monitoring system, in *16th International Conference on ICT and Knowledge Engineering (ICT&KE), Bangkok*, pp. 1–4 (2018)
6. K. Venkatesh, P. Rajkumar, et al., IoT based home automation using Raspberry Pi. J. Adv. Res. Dyn. Control Syst. **10**(07-Special Issue) (2018)
7. E. Rammohana Reddy, K. Sankara, Internet of Things based home automation control system using Raspberry Pi. Int. J. Sci. Res. Comput. Sci. Eng. Inf. Technol. (2018)
8. B. Mataloto, J.C. Ferreira, N. Cruz, IoT for building and energy management systems. J. Electron. **8**(7), 763–770 (2019)
9. P. Patel, H. Patel, Various technologies using for smart home automation based on Internet of Things. J. Res. **05**(2) (2019)
10. J. Naik, K. Narare, S. Patil, P. Salve, A.A. Bamanikar, Home automation system using IOT. Int. Res. J. Eng. Technol. (IRJET) **06**(22) (2019)

Chapter 13
Modeling and Simulation of Solar Photovoltaic Array Using MATLAB

Vagish Kumar and Rajeev Kumar

Abstract The solar energy plays an important role in the accomplishment of the renewable energy goals of the world. The modeling and simulation are a vital part of the analysis of the PV system before its installation. It provides the understanding of the behavior of the system under the actual working condition in priori. In this paper, a simulation of the solar PV module using the MATLAB is presented. The mathematical expressions based on the theory of semiconductor are used. The simplified solar PV cell equivalent circuit is used for the analysis. The expression obtained for the ideal cell is modified to take into account the shunt and the series resistances due to the surface effect and the presence of various connections. The objective is to obtain the *I–V* and the *P–V* characteristics of the solar PV cell to estimate its performance under the various operating conditions. The simulation is conducted for the three most important parameters, namely operating temperatures, intensity of irradiance and the shunt resistance, and the results are analyzed.

Keywords Solar energy · Solar PV cell · PV module · MATLAB

13.1 Introduction

Over the past decade, the role of the renewable energy in meeting the growing demand of the energy has increased manifold. The issue of the global warming is gaining prominence day by day, and the use of the solar energy assists in not only reducing the greenhouse gas emission, but it also provides the flexibility to achieve the desired energy mix [1]. It reduces the dependence on the fossil fuel. For countries like India where the major chunk of the fossil fuel demand is met through the import. Even if a small portion of the total solar energy incident on the surface of the earth is converted into the useful energy, the problem of the energy crisis and its adverse effect [2] can be eliminated.

V. Kumar (✉) · R. Kumar
Indian Institute of Technology Mandi, Mandi, Himachal Pradesh 175005, India
e-mail: T18127@students.iitmandi.ac.in

R. Seyezhai et al. (eds.), *Recent Trends in Renewable Energy Sources and Power Conversion*, Springer Proceedings in Energy,
https://doi.org/10.1007/978-981-16-0669-4_13

The solar energy is becoming more affordable day by day due to the improvement in the performance and the reduction of the price because of mass production. In India, the focus has been lately on the wind and the solar energy. Thus, the understanding of the performance of the solar cell under various working conditions is of paramount importance.

13.1.1 Working of Solar PV Cell

The photoelectric effect forms the basis of the working of the solar PV cell. It is a phenomenon of the ejection of the electron from the valence band when the material is subjected to the sunlight of a particular wavelength which depends on the band gap. It basically consists of the p–n junction which is exposed to the sunlight present in the environment. The p-type and the n-type semiconductors contain the holes and the electrons, respectively, in excess. When the p-type semiconductor and the n-type semiconductor are brought in contact of each other, a depletion zone is formed at the boundary [3] which leads to the formation of the junction.

When the sunlight falls on solar cell's surface, a fraction of it is absorbed by the material. When the energy content of the incident sunlight is higher than the semiconductor material's band gap of which the solar cell is made, then the electron is ejected from the valence band and gets transferred to the conduction band (C.B). Due to this, the electron–hole pair comes into existence. The electron which reaches the C.B is free to move. The electric field which is produced due to the formation of the junction forces the free electron in the C.B to move in the specified direction. This leads to the production of the currents. The electrodes are connected to the top and the bottom surface of the cell to connect it to the external load. The solar cell produces DC current but most of the loads require AC current. The inverter is used to convert DC to AC. The inverters used in the solar applications have high conversion efficiency of around 90% and above. Apart from conversion, it also acts as the safety valve, i.e., it disconnects the solar PV system from the grid, whenever there is a power loss in the grid.

A typical solar PV system consists of circuit breakers, combiners, meters, disconnectors and wirings. The disconnector is present on both sides of the inverters, namely AC and DC. It provides insulation when the inverter has to be installed or repaired. The circuit breaker safeguards the system when there is sudden surge in the current in the system. There are both manual as well as automatic circuit breaker. The automatic circuit breaker disconnects the system automatically once it reaches the preset value. The electric meter measures the energy coming in and going out of the system. The solar PV system connected to grid usually consists of bidirectional meter which measures both the energy flow to and from the grid. The wiring is done to connect the various components of the system. The wiring of the solar PV system is exposed to the sunlight for the prolonged period of time, so it requires special protection to save it from the UV rays.

13.1.2 Literature Review

Tsai et al. [5] proposed a model to obtain the I–V and P–V characteristics of PV module. They observed that as there is increase in the solar irradiance's intensity, the short-circuit current as well as maximum output power increases, whereas with rise of solar cell's temperature, Isc increases but maximum output power decreases. The P–V characteristic of a sample day of operating temperature 25 °C and irradiance of 1 kW/m^2 was obtained. The modeling was done using Simulink. Pandirajan et al. [6] demonstrated that with increases of ambient temperature, the output current did not show any appreciable change, but drastic decrease in the output voltage was observed. It led to the reduction in net power output with the increase of temperature. The simulation was performed for 25 °C, 50 °C and 75 °C, and power output obtained at the constant irradiance was 38 W, 33 W and 29 W, respectively. Keles et al. [7] demonstrated a Simulink model of the solar PV module which is supplying current to the external DC load. The reference temperature and the irradiance were taken as 25 °C and 1000 W/m^2, respectively. The transient response of the system under the DC load was obtained. It was found that the response time can be adjusted by changing the value of the internal resistance and the capacitance. The time taken to attain the steady-state response was investigated. Dhar et al. [8] proposed a circuit-based model of a PV cell to find the output of PV cell w.r.t the changing environmental conditions like solar irradiance's intensity and cell's temperature. It was demonstrated that with increase in irradiance, the power output of solar panel increases. Dev et al. [9] did the modeling of photovoltaic module. It took irradiance and temperature as variable input parameters and obtained the characteristic plot. It observed that there is a point at which the output power becomes maximum for a given value of parameters. Since the PV module possess nonlinear characteristics, it plays a vital role in the design of the system to track these points for solar PV applications. Abdullahi et al. [10] modeled monocrystalline PV module using MATLAB/Simulink and compared the experimental and the simulation results. The results match closely, but it does not discuss the various steps involved in the modeling in detail. Pendem et al. [11] performed KC200GT PV cell's modeling in MATLAB/Simulink and also presented the effect of partial shading on the module's characteristics but lacked the stepwise description of the modeling. Thus, it becomes difficult for the researchers to understand the model.

Since the solar energy has huge impact on the energy security of any country, so it can be said that it is vital for the development. Thus, it is necessary to understand the performance of the system under various real working conditions using the simulation tools. There are many software packages available for the modeling of the PV module, but they are only commercially available and expensive. This problem can be resolved by using MATLAB/Simulink for the modeling. It gives accurate result, and also it is user friendly and easily available. In this paper, the author proposes the analytical method using MATLAB/Simulink to develop the single diode model of the solar module. The stepwise modeling is proposed firstly for the ideal solar cell and then for the real solar cell by taking into account the parallel and the series resistance.

13.2 Mathematical Modeling

The solar PV module comprises the number series and the parallel connected cells. The arrays are formed by further connecting these modules in parallel and series to obtain the power. The solar PV characteristics have three crucial points, namely open-circuit voltage, short-circuit voltage and the maximum power point. The value of these parameters can be found from the data sheets for the given module.

The solar PV cell is represented by a current source which has parallel connection with the diode as shown in Fig. 13.1. Kirchhoff's law can be applied to obtain the output current of the cell as shown in Eq. (13.1)

$$I = I_{ph} - I_d \tag{13.1}$$

The I–V characteristics of the solar PV cell are given by Schockley's equation. It is obtained from the theory of semiconductors as illustrated in Eq. (13.2)

$$I_{\mathrm{d}} = I_{\mathrm{rs}}\left[\exp\left(\frac{qV_{\mathrm{oc}}}{N_{\mathrm{s}}\mathrm{KAT_o}}\right) - 1\right] \tag{13.2}$$

By substituting the expression of the diode current (), the output current is obtained as shown in Eq. (13.3)

$$I = I_{\mathrm{ph}} - I_{\mathrm{rs}}\left[\exp\left(\frac{qV_{\mathrm{oc}}}{N_{\mathrm{s}}\mathrm{KAT_o}}\right) - 1\right] \tag{13.3}$$

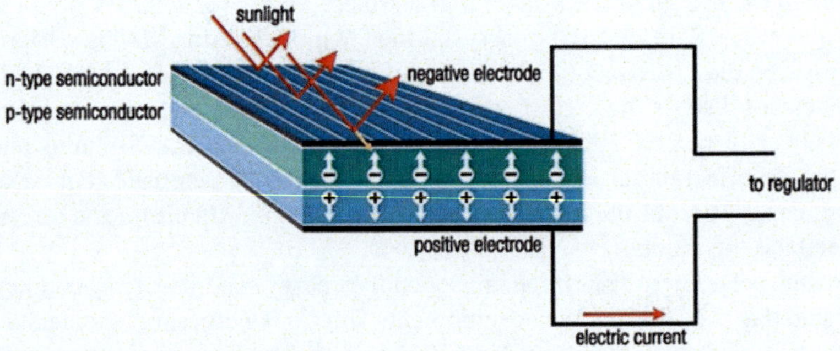

Fig. 13.1 Cross section of the solar PV cell [4]

13.2.1 Modeling of Real Solar PV Cell

In the ideal solar PV cell model, the current output was assumed to be directly proportional to the solar irradiance. There are various other parameters which affect the output current of the solar PV cell. The series and the parallel resistance are introduced in the real solar cell model as shown in Fig. 13.2. These were not considered in the ideal model, but it cannot be ignored because it significantly affects the efficiency of the solar PV cell. The series resistance in the circuit is represented by R_s. It takes into account the resistance due to grids, collecting bus, semiconductor materials and the connections. The parallel resistance, R_p, is also called as shunt resistance which takes into account the leakage of current because of the surface effects and the thickness of the cell. The value of the R_p is significant only when the large number of modules is connected together. Therefore, the R_p is taken as infinite, and only R_s is considered further (Fig. 13.3).

Considering the R_s, Eq. (13.3) is modified as shown in Eq. (13.4)

$$I = I_{\mathrm{ph}} - I_{\mathrm{rs}}\left[\exp\left(\frac{q(V + \mathrm{IR_s})}{N_s \mathrm{KAT_o}}\right) - 1\right] \tag{13.4}$$

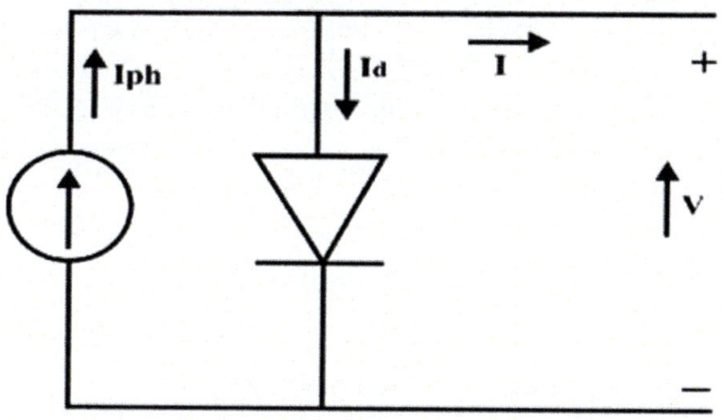

Fig. 13.2 Ideal solar PV module circuit

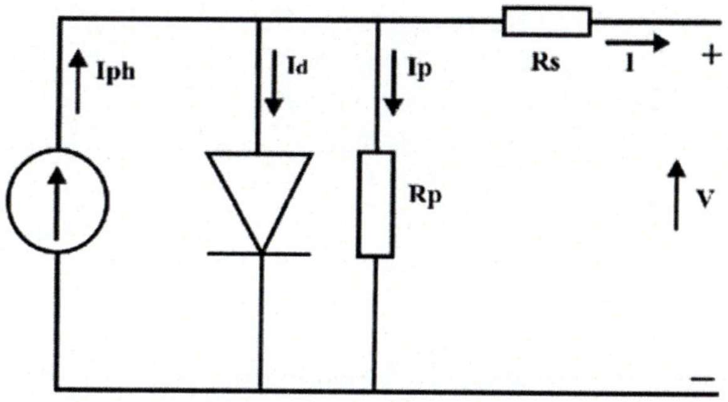

Fig. 13.3 Real solar PV cell equivalent circuit [12]

When the coupling of the solar PV cell in the series–parallel configuration is taken into account, the output current is given as

$$I = N_p * I_{ph} - N_p * I_s\left[\exp\left(\frac{q(V + IR_s}{N_s KAT_o}\right) - 1\right] \tag{13.5}$$

In the above presented model of solar PV cell, it is required to determine some parameters like Iph, Irs and Is which depend on the PV module used for the analysis. The proto current has linear dependence on the solar irradiation to which the module is exposed. The expression for the photocurrent is given by Eq. (13.6)

$$I_{ph} = [I_{sc} + Ki(T_o - T_i)] * \frac{G}{G_{ref}} \tag{13.6}$$

The reverse saturation current and the saturation current are given by Eqs. (13.7) and (13.8)

$$I_{rs} = I_{sc}/\left[\exp\left(\frac{qV_{oc}}{N_s KAT_o}\right) - 1\right] \tag{13.7}$$

$$I_s = I_{rc}\left[\frac{T_o}{T_r}\right]^3 \exp\left[\left(\frac{qE_g}{AK}\right)\left(\frac{1}{T_r} - \frac{1}{T_o}\right)\right] \tag{13.8}$$

The output power is obtained given in Eq. (13.9) (Table 13.1)

$$P = I * V \tag{13.9}$$

where

A = Ideality factor.
I_d = Diode current [A].
N_s = Number of series connected cells.
N_p = Number of parallelly connected cells.
q = Charge of an electron.
T_o = Operating temperature [K].
T_r = Reference temperature [K].
I_{sc} = Short-circuit current at STC [A].
K = Boltzmann's constant.
I_{ph} = Current produced because of photoelectric effect [A].
I_{rs} = Reverse saturation current of the diode [A].
G_{ref} = Reference value of irradiance's intensity [W/m^2].
G = Solar irradiance's intensity [W/m^2].
K_i = Coefficient of temperature.
E_g = Semiconductor's energy band gap.
V_{oc} = Open-circuit voltage [V].
R_s = PV module's series resistance [Ω].

Table 13.1 Values of constants used in the simulation

Sl. No.	Parameters	Value
1	T_r	298.15 K
2	Gref	1000 W/m^2
3	q	1.6e−19 C
4	k	1.38e−23 J/K
5	Eg	1.115 eV
6	I_{sc}	3.6 A
7	V_{oc}	0.8 V
8	A	2
9	K_i	0.005254 A/°C

R_p = PV module's parallel resistance [Ω].
V = PV module's voltage output [V].
I = PV module's current output [A].

13.3 Simulation Results and Discussion

The performance of the solar PV cell is influenced by various parameters. The value of these parameters does not remain constant. It varies with the time of the day, seasons, weather and various other factors. The output characteristics of the solar PV cell were investigated for the various temperatures and irradiances. The program was written using the MATLAB based on the expressions developed in the previous section. The simulation was performed for varying parameters by employing loop.

13.3.1 Influence of Operating Temperature

The investigation was done to analyze the effect of ambient temperature on the solar PV module's power output. The intensity of irradiance was taken to be constant value of 1 Kw/m^2. Then simulation was performed for the temperature of 40, 55 and 70 °C as shown in Figs. 13.4 and 13.5.

It was observed that the temperature influences the solar PV cell's performance appreciably. As the temperature increases, the output power decreases. So, to increase the PV cell's efficiency, the temperature of the module should be maintained low. There are various cooling methods to achieve this.

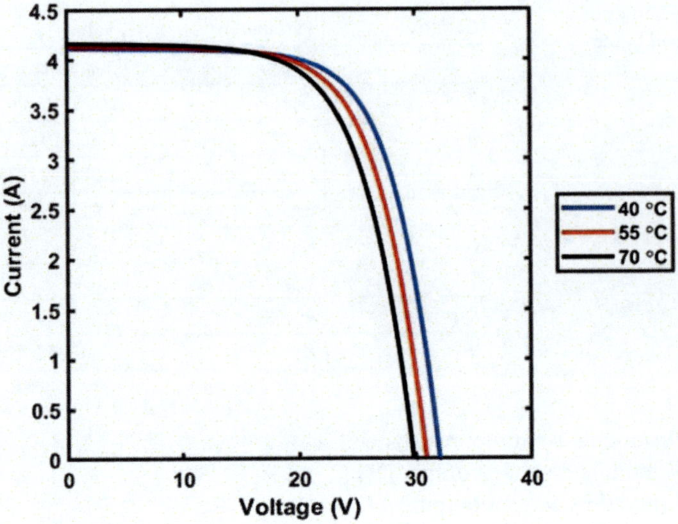

Fig. 13.4 *I–V* characteristics at variable temperature

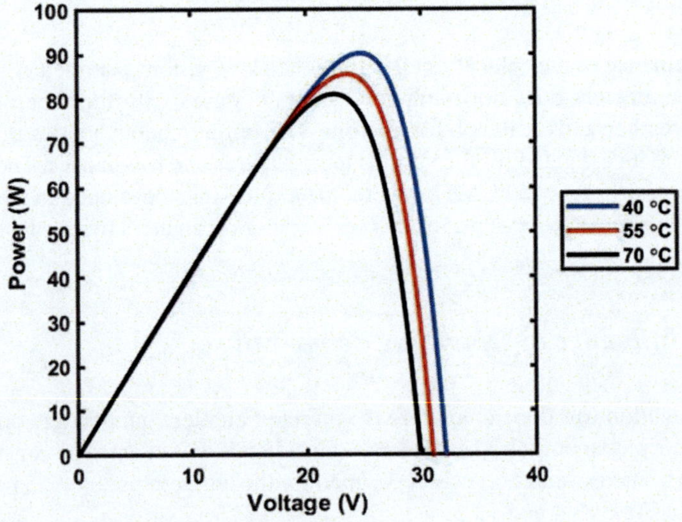

Fig. 13.5 *P–V* characteristics at variable temperature

13.3.2 Influence of Irradiance Intensity

The simulation was conducted to analyze the effect of intensity of solar irradiance on the *I–V* and the *P–V* characteristics as shown in Figs. 13.6 and 13.7. The temperature

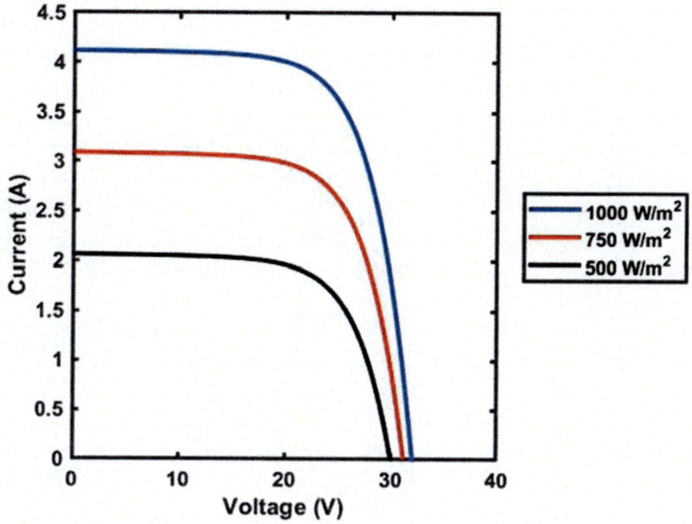

Fig. 13.6 *I–V* characteristics for varying irradiance

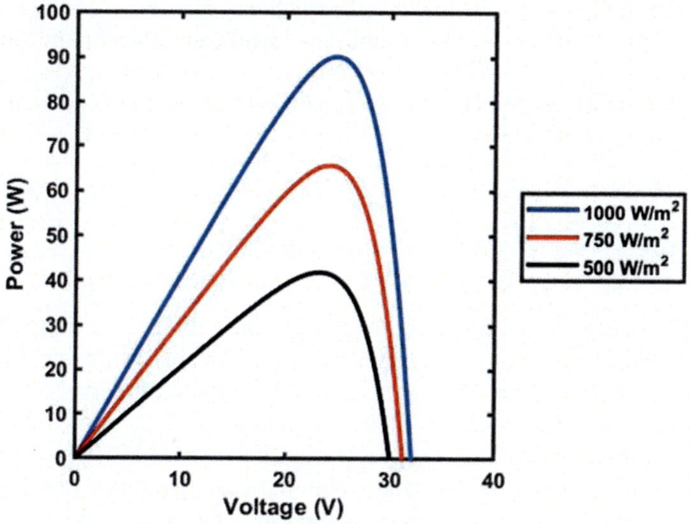

Fig. 13.7 *P–V* characteristics for varying irradiance

of the solar PV module was kept constant at 40 °C, and the investigation was done for the irradiance of 500, 750 and 1000 W/m^2.

It was observed that as the solar irradiance's intensity increases, the increase in the power output of the solar PV module also happened. The output current remains constant as the voltage increases till 22.6 V, and after that it starts decreasing. It can

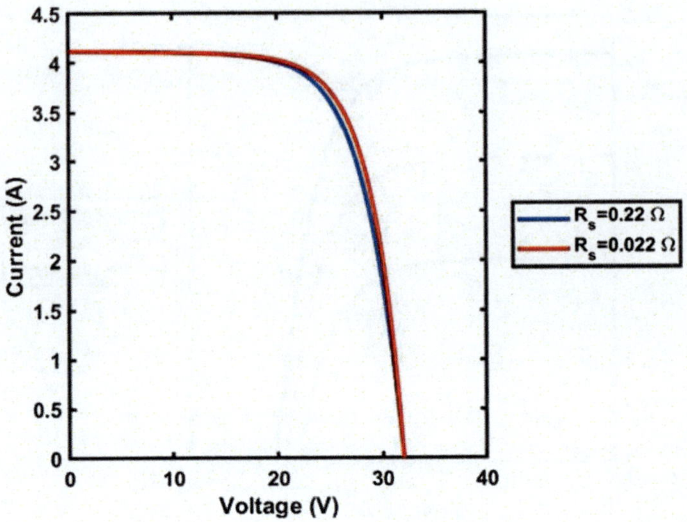

Fig. 13.8 Influence of series resistance on *I*–*V* characteristics

be inferred from Figs. 13.6 and 13.7 that the irradiance has appreciable impact on the short-circuit current where it did not show any significant effect on the short-circuit voltage.

The irradiance's intensity varies with the time of the day, weather condition, cloud cover and various other factors.

13.3.3 Influence of Series Resistance

The simulation was carried out to obtain the influence of the series resistance on the solar PV module's *I*–*V* and *P*–*V* characteristics. The temperature and the intensity of irradiance were kept constant at 40 C and 1000 W/m^2, respectively. The analysis was done for the series resistance of 0.22 Ω and 0.022 Ω as shown in Figs. 13.8 and 13.9. It was observed that the series resistance does not have appreciable impact on the short-circuit voltage and current. With decrease in value of the resistance in series, the value of the output power increases but not substantially.

13.4 Conclusion

This paper presents a solar PV cell modeling using MATLAB. It analyzes the effect of temperature, irradiance and the series resistance on the *I*–*V* and *P*–*V* characteristics of the solar cell. The *I*–*V* characteristics show that the power output attains the

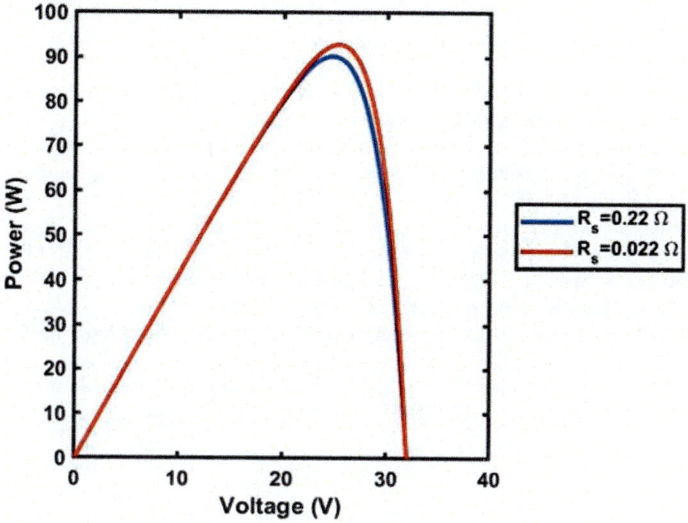

Fig. 13.9 Effect of series resistance on *P*–V characteristics

maximum value corresponding to a particular operating condition. This model is of paramount importance in the solar PV applications for the purpose of obtaining the point at which maximum power output occurs. The proposed model is very useful for the research in the field of solar PV cell's performance enhancement. In the future, the solar energy is going to play crucial role in solving the issues of energy crisis and the harmful environmental effects associated with the use of fossil fuels.

References

1. S. Kanchan, A. Kumarankandath, The Indian power sector: need of sustainable energy access. IPPAI Knowledge Report, pp. 126–140. https://cdn.cseindia.org/attachments/0.61046700_1505884602_The-Indian-Power-Sector.pdf (2015)
2. K. Shahzad Baig, M. Yousaf, Coal-fired power plants: Emission problems and controlling techniques. J. Earth Sci. Clim. Change **8**(404), 1–9. https://doi.org/10.4172/2157-7617.1000404
3. N.K. Manoj, S. Jaswant, S. Baljit, S. Silpa, S. Vindhu, T. Sachin, Introduction to semiconductor nanomaterial and its optical and electronics properties. Micro Nano Technol. 1–33 (2017). https://doi.org/10.1016/B978-0-323-44922-9.00001-6
4. Revision Energy. (2018). Last assessed 28th July 2020. https://www.revisionenergy.com/solar-power-for-your-home/all-about-solar-inverters/
5. H.-L. Tsai, S. Tu, Development of generalised photovoltaic model using MATLAB/SIMULINK, in *Proceedings of WCECS* (2008)
6. N. Pandirajan, R. Muthu, *Mathematical Modeling of Photovoltaic Module with Simulink* (Department of Electrical and Electronics Engineering, SSN college of Engineering, 2011). https://doi.org/10.1109/ICEES.2011.5725339

7. K. Cemal, A.B. Baykant, A. Murat, K. Ashim, Photovoltaic system model for Matlab/simulink simulation, in *4th International Conference on Power Engineering, Energy and Electrical Drives* (2013). https://doi.org/10.1109/PowerEng.2013.6635863
8. S. Dhar, R. Sridhar, *Modelling and Simulation of Photovoltaic Arrays* (Department of Electrical and Electronics Engineering, SRM University, 2012)
9. A. Dev, S. Berclin Jeyaprabha, Modelling and simulation of photovoltaic module in MATLAB, in *Proceedings of the International conference on Applied Mathematics and Theoretical Computer Science* (2013)
10. N. Abdullahi, C. Saha, R. Jinks, Modeling and performance analysis of a silicon PV module. J. Renew. Sustain. Energy **9**(3), 1–11 (2017). https://doi.org/10.1063/1.4982744
11. S.R. Pendem, S. Mikkili, Modeling, simulation and performance analysis of solar PV array configurations (Series, Series-Parallel and Honey-Comb) to extract maximum power under Partial Shading Conditions. Energy Rep. **4**, 274–287 (2018). https://doi.org/10.1016/j.egyr.2018.03.003
12. N. Mendalek, K. Al-Haddad, Photovoltaic system modeling and simulation, in *Industrial Technology, ICIT* (IEEE, pp. 1522–1527, 2017). https://doi.org/10.1109/ICIT.2017.791592

Chapter 14
Improvement of Power System Stability and Power Quality of Artificial Intelligent Controller Based Grid-Connected PV System Using Cascaded Multilevel Inverter

I. Mahendravarman and M. Venkateshkumar

Abstract In the present scenario, the photovoltaic (PV)-based power generation is widespread in numerous countries due to friendly. The power electronics converters are playing a vital role in the PV power conversion system. Presently boost and buck-boost converters are used with the MPPT algorithm in the PV system, which has less conversion efficiency and also in a grid-integrated PV system to maintain the good power quality is a major problem. A fuzzy—MPPT controller was designed to control a super lift Luo converter—based PV system, and the super lift Luo converter was studied and simulated at different topologies and weather conditions to improve the PV system's conversion efficiency, and a new Cascaded Multi Level Inverter (CMLI) with fuzzy controller is proposed here to minimize THD and achieve good power quality at PV—Grid integration. The new CMLI comes with a minimal range of solid-state switches and is most versatile for the PV system and can be worked on asynchronous voltage sources. The output standard of the proposed CMLI had improved in which total harmonic distortion (THD) is minimized and the PV system's power efficiency had also improved. In a broad range of operating conditions, this theoretical PV model simulated under the MATLAB environment and its performance was analyzed. The proposed converter and MPPT algorithm are used by the artificial intelligent controller to grid integration of the PV system by way of a new CMLI to increase power quality and reliability. Finally, the simulation results of the proposed system are evaluated and verified with the IEEE 519 and 1547 standard for proving the usefulness of the proposed system.

Keywords Photovoltaic · MPPT · Fuzzy · Cascaded multilevel inverter · THD

I. Mahendravarman (✉)
A.V.C College of Engineering, Mayiladuthurai, India

M. Venkateshkumar
Amrita School of Engineering, Amrita Vishwa Vidyapeetham, Chennai, India
e-mail: m_venkateshkumar@ch.amrita.edu

R. Seyezhai et al. (eds.), *Recent Trends in Renewable Energy Sources and Power Conversion*, Springer Proceedings in Energy,
https://doi.org/10.1007/978-981-16-0669-4_14

14.1 Introduction

Nowadays, the world facing the foremost problem is global warming due to greenhouse gas as well as growing temperature day by day due to fossil fuel-based power generation and the use of transport vehicles. Many countries have now focused on alternative renewable energy sources such as PV, wind, fuel cell, hydro, and geothermal. In January 2011, the MNRE of India amended the regulation to prescribing solar-specific RPOs from at least 0.25% in 2012 to 3% by 2022. Furthermore, the NAPCC suggests that the share of clean energy in the overall energy mix will rise to a minimum of 15% by 2020. In the distribution of renewables and electricity supplies, which have historically been commonly used in various types of network structures, technological development in solid-state electronics plays a significant role. Similarly, multiple checking methodologies and interface elements were applied and implemented. They primarily decide to add to the benefits of the system when satisfying the load needs by configuring their devices and control techniques. Renewable and substitute energy supplies are commonly scattered in microgrids in current scenarios. These technologies are favored in the context of a microgrid because of various benefits such as the best utilization of energy resources, improved efficiency of electricity, and increased stability of the source. Newly, more sophisticated grid infrastructure has evolved with the grid area infrastructure, multi-microgrids, interconnected AC-AC microgrids, and interconnected AC-DC microgrids. This innovative network architecture aims specifically to optimize clean energies and sustainable energy resources [1]. For instance, it allows reserve sharing, resistance voltage, and frequencies by interconnecting two or more microgrids and eventually enhances the overall efficiency and durability of interconnected microgrids [2]. PV and fuel system clean sources of electricity-producing DC outputs. In the configuration and management of the output voltage, the conversion of DC-DC and AC-DC converters plays an important role. Diverse types of converters are currently being used for DC-DC and MPPT converters such as boost, buck, buck–boost, and Cuk converters. The main drawbacks of the conventional controllers are producing high ripple voltage and current at the output. In the last decade, many researchers focused on the above-said issue and find the solution to overcome by super-lift Luo converter [3]. The advantages of the super-lift Luo converter are high voltage transfer gain, high power density, strong efficacy, decreased ripple voltage, and current. Therefore, the super-lift Luo converter is more suitable for a photovoltaic system to boost the voltage as well as MPPT controller design [4].

Design and analysis of the fuzzy controller-based MPPT algorithm for 5300 W PV panel using super-lift Luo converter are discussed in Sect. 14.2. Then design and analysis of 13 levels of new CMLI for the PV system are discussed in Sect. 14.3. Further in Sect. 14.4, the design and analysis of grid interfacing of the photovoltaic system via new CMLI and controlled by a fuzzy logic algorithm are discussed and finally, the conclusion was discussed in Sect. 14.5.

14.2 Fuzzy-MPPT-Based PV System

14.2.1 The Framework of Photovoltaic System

Renewables focus solely on natural resources contributing to nonlinear power generation. The MPPT algorithm is important for controlling and produce full power under different operating conditions for all renewable energies [4]. The MPPT algorithm is important. Conventional controls for MPPT algorithms like P&O, incremental conductance, voltage and current, input capacity, and some systems-based expert algorithms have been used in recent years. Due to the low voltage generation of a PV cell, a single PV cell cannot satisfy the market demand for electricity. PV cells are thus organized as a high energy series.

This proposed system is designed with super-lift Luo converter and presented Fig. 14.1. In this circuit topology, we can connect solar array at input side instead of DC voltage source. In solar array, the PV cells were first ordered in series to attain the desired voltages and then organized in parallel so that more current can be generated. A 5300 W PV module was planned here in this way. For the 5300 W PV system, the fuzzy-MPPT algorithm based on a super-lift Luo converter was built in this text, as shown in Fig. 14.2, where two input sources are available, such as solar irradiance and temperature.

The 5300 W PV array is attached to the super-lift Luo converter with a fuzzy MPPT-based controller and can generate a gate signal for the converter switches. The output of the PV device was supplied to the fuzzy-MPPT controller [5] for analysis and produces a reliable PV output that adapts to different irradiance conditions.

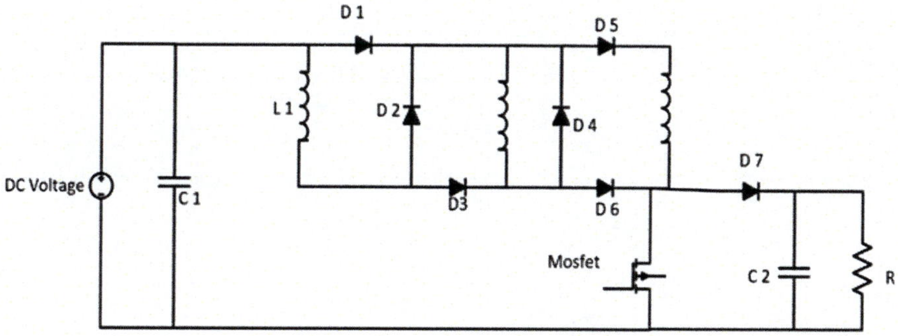

Fig. 14.1 Super-lift Luo converter topology

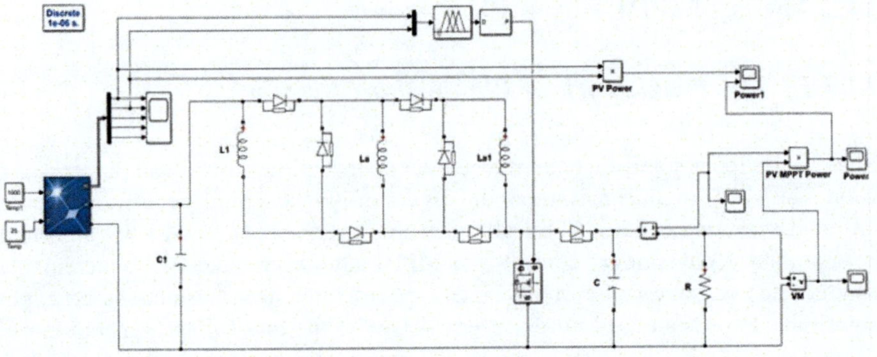

Fig. 14.2 PV-connected super-lift Luo converter with fuzzy-MPPT controller

14.2.2 *Fuzzy-Based MPPT Controller*

There are two inputs of the fuzzy controller which are PV voltage and current with one output, which is the duty cycle of solid-state switches in converter as seen in Fig. 14.3.

The fuzzy input and output membership function has been developed by using the trapezoidal function and each membership function is classified into three ranges such as low, medium, and high. Based on these ranges, fuzzy rules have been developed for input and output membership functions as mentioned in Table 14.1.

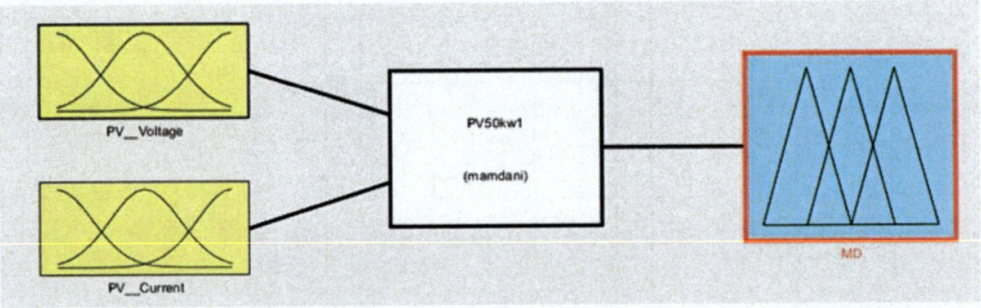

Fig. 14.3 Fuzzy controller-based MPPT algorithm

Table 14.1 Fuzzy logic for the proposed system

PV panel voltage	Low	Medium	High
Low	High	High	Medium
Medium	High	Medium	Medium
High	Medium	Medium	Low

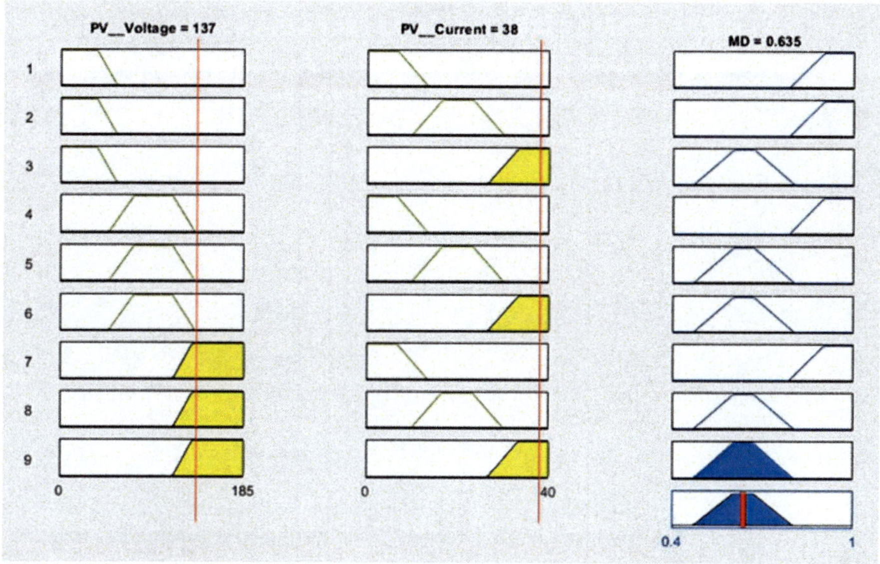

Fig. 14.4 Fuzzy rules-based system for MPPT algorithm

In this case, the centroid process was used to transform the fuzzy value into a crisp value [6]. The fuzzy intervene rules are drawn up according to PV module input and output variables presented in Fig. 14.4.

14.2.3 PV System Outputs

Two-stage super-lift Luo converter

The PV system from the super-lift Luo converter generates higher power than the PV system from the boost converter and is shown as a waveform for power comparison in Fig. 14.5. Here, the super-lift Luo converter-based PV system generates 5300 W of converter-side power while the booster-based PV system produces 5000 W of converter-side power. Normally boost converter is used in PV-based power systems but has less conversion efficiency.

The MPPT algorithm for the PV system is designed by a fuzzy controller which generating the duty cycle for the super-lift Luo converter (two-stage) existing in the PV system. In which, PV system performances such that power, voltage, and current have been analyzed under various weather conditions. Here, PV input power and PV-MPPT power have been compared and presented in Fig. 14.6.

Here, the super-lift Luo converter-based PV system gain was even higher than the boost converter-based PV system. The voltage and current waveforms are shown in

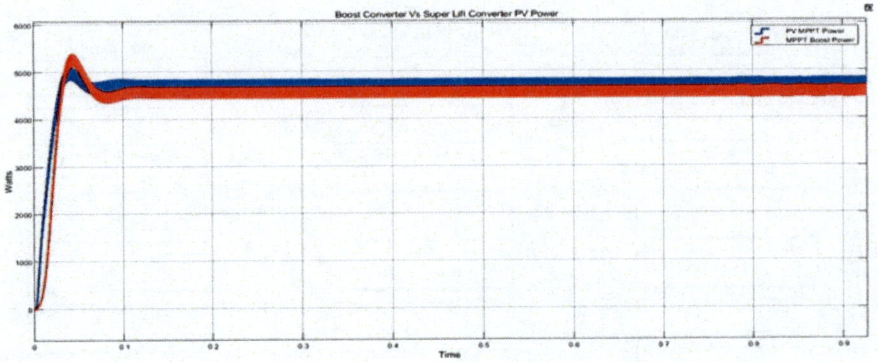

Fig. 14.5 Comparative analysis of boost and super-lift converter for 5300 W PV system

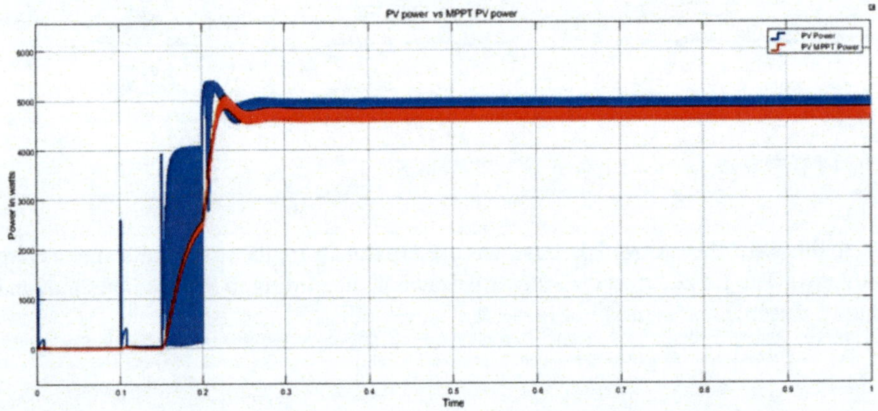

Fig. 14.6 Two-stage super-lift Luo converter power waveform (PV and PV-MPPT)

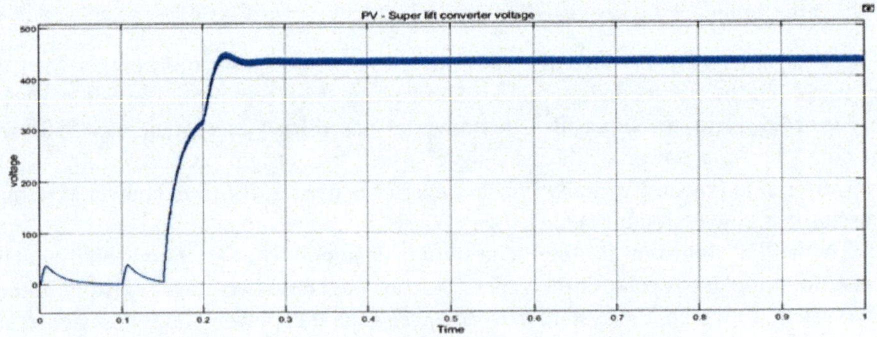

Fig. 14.7 Two-stage super-lift Luo converter voltage waveform (PV MPPT)

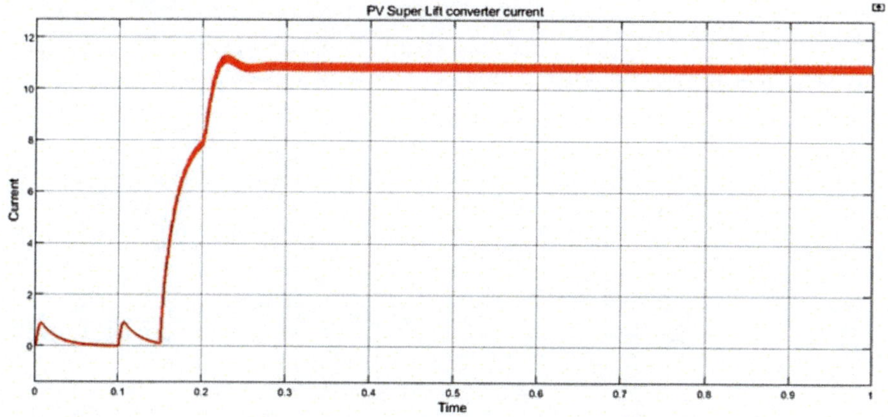

Fig. 14.8 Two-stage super-lift Luo converter current waveform (PV-MPPT)

Figs. 14.7 and 14.8, respectively, and the voltage was improved even higher than the traditional converter.

Three-stage super-lift Luo converter

The three-stage Luo converter is created, and the above analyses are performed and the performance features are analyzed under various weather conditions. Here, the voltage and current waveforms are presented in Figs. 14.9 and 14.10, respectively. The voltage level here was much higher than the two-stage Luo converter.

Three-stage super-lift Luo PV input power and PV-MPPT power have been compared and presented in Fig. 14.11. Based on the analysis, the three-stage super-lift Luo

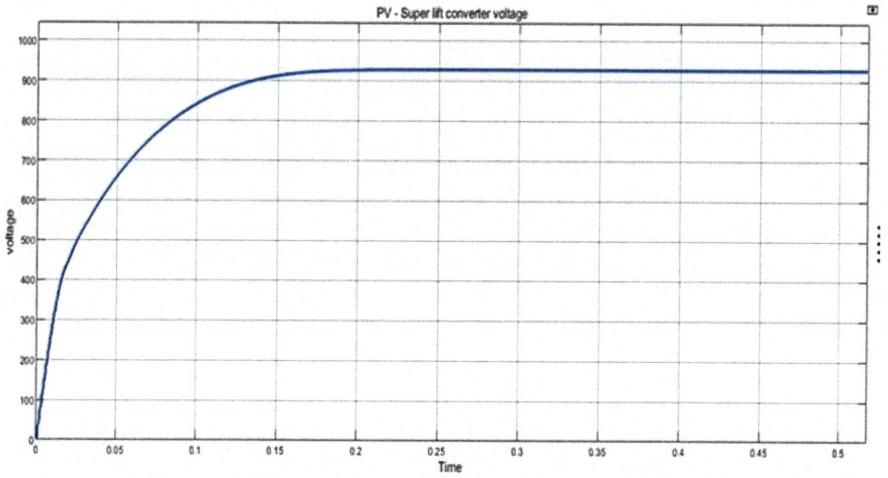

Fig. 14.9 Three-stage super-lift Luo converter voltage waveform (PV-MPPT)

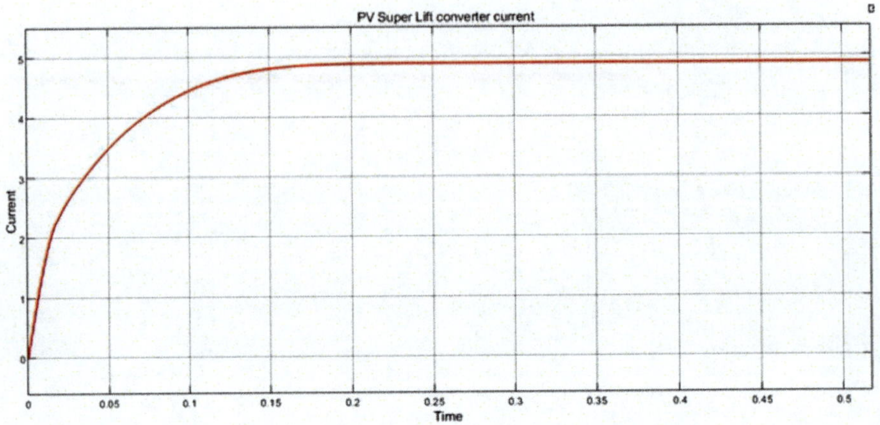

Fig. 14.10 Three-stage super-lift Luo converter current waveform (PV-MPPT)

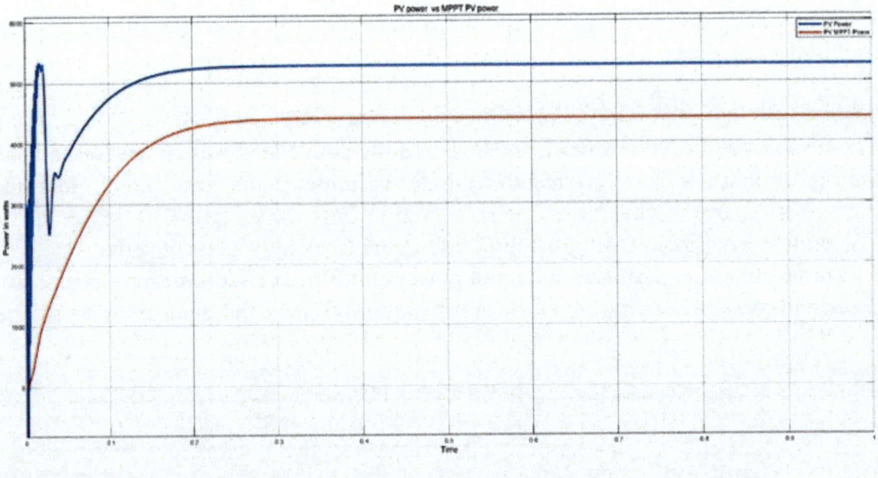

Fig. 14.11 Three-stage super-lift Luo converter power waveform (PV and PV-MPPT)

converter voltage conversion gain and reliability have been higher than the two-stage converter and traditional converter.

It is more suitable for the PV-MPPT system where three-stage super-Luo converters are installed and used for the grid-connected system under typical irradiation.

14.3 Proposed Cascaded Multilevel Inverter

The CMLI is best suited for the green energy grid integration in order to increase the system energy efficiency and reliability [7–9]. In order to lower the THD level for the current and voltage waveform, the proposed CMLI was built with a minimal number of solid-state switches.

Six PV sources for grid integration to produce 32 kW of power where each PV source has a single solid-state switch and CMLI configuration has six solid-state switches as shown in Fig. 14.12 [6, 10].

The modular staircase technique is used to produce the pulses for the proposed CMLI [5]. This new topology of CMLI is built and simulated in the MATLAB tool and shown in Fig. 14.13. The voltage source controller is used to interface with the grid.

Simulation has been performed and the proposed method had produced pure sinusoidal output at the CMLI terminal. The three-phase output voltage of the device is 440 V and is shown in Fig. 14.14. This inverter design increases the performance of the device By lowering the level of THD for both current and voltage waveforms.

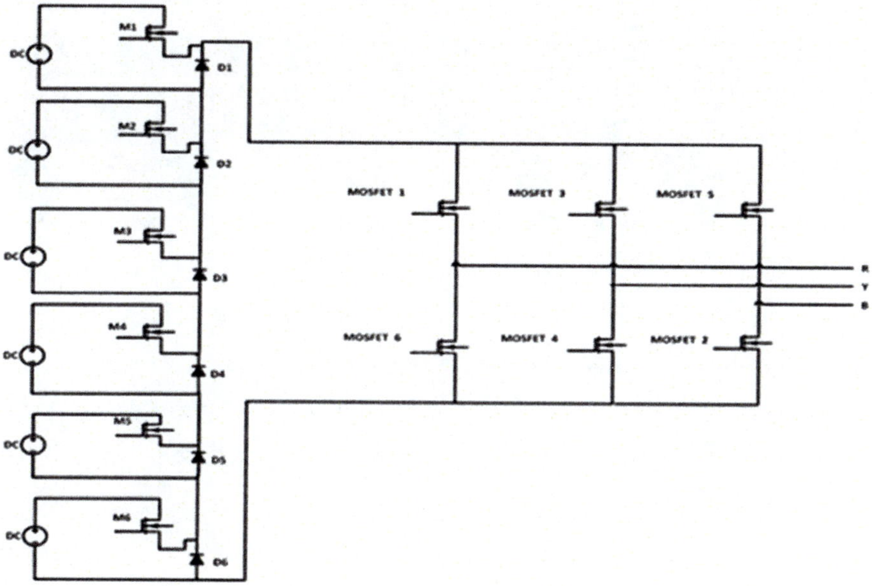

Fig. 14.12 New CMLI topology for grid integration PV system

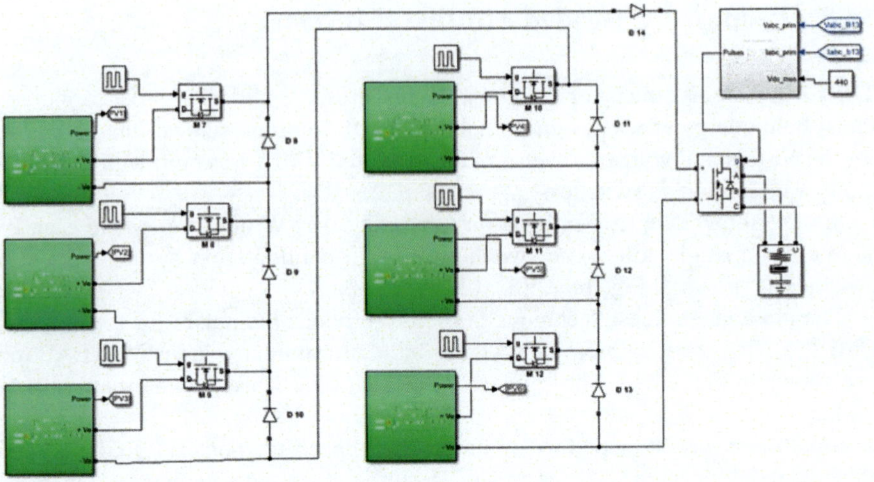

Fig. 14.13 MATLAB/Simulink model for proposed system

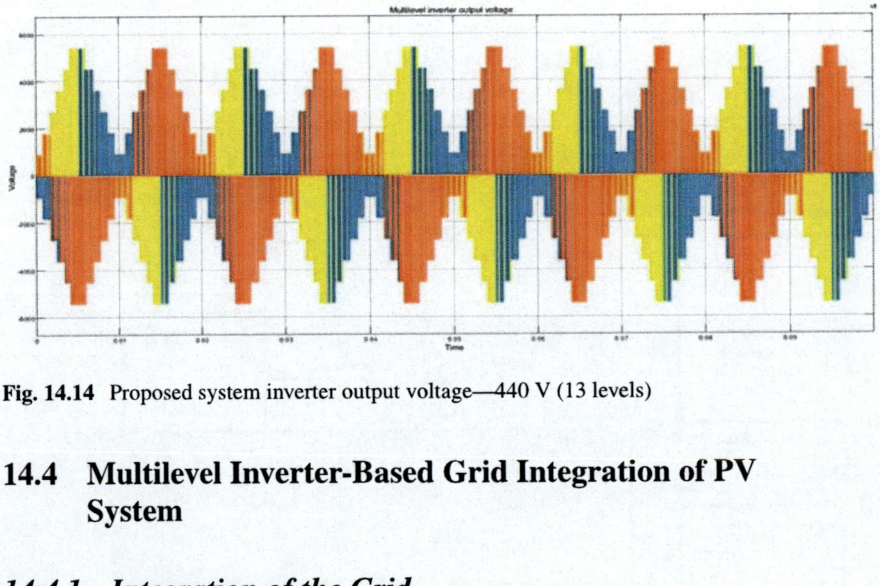

Fig. 14.14 Proposed system inverter output voltage—440 V (13 levels)

14.4 Multilevel Inverter-Based Grid Integration of PV System

14.4.1 Integration of the Grid

The grid-interfaced proposed PV system configuration is seen in Fig. 14.15 and the grid integration model of the PV system using the multilevel inverter proposed to be operated by an artificial intelligent controller as seen in Fig. 14.16. With the aid of the new cascade multilevel inverter, six PV sources are installed in the grid.

The inverter provides the pure grid output, while grid parameters will be taken to generate PWM for the proposed MLI as a reference for the voltage source converter

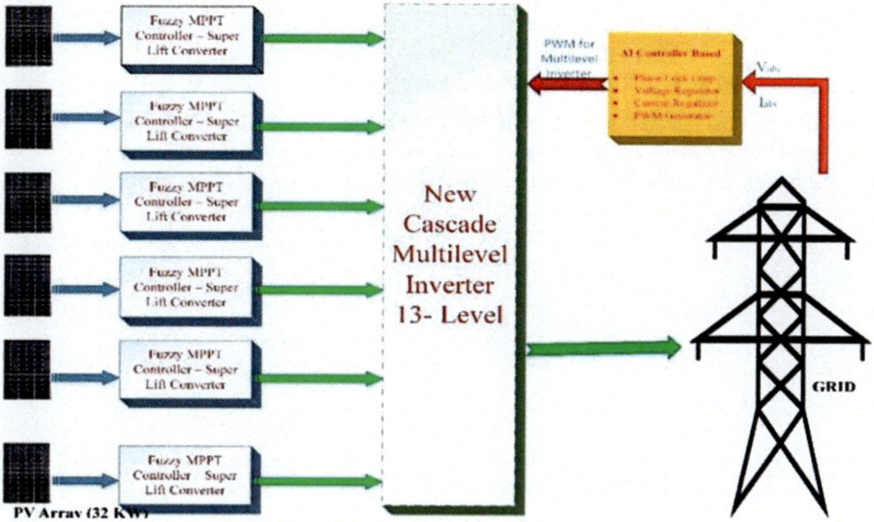

Fig. 14.15 Layout of the proposed system

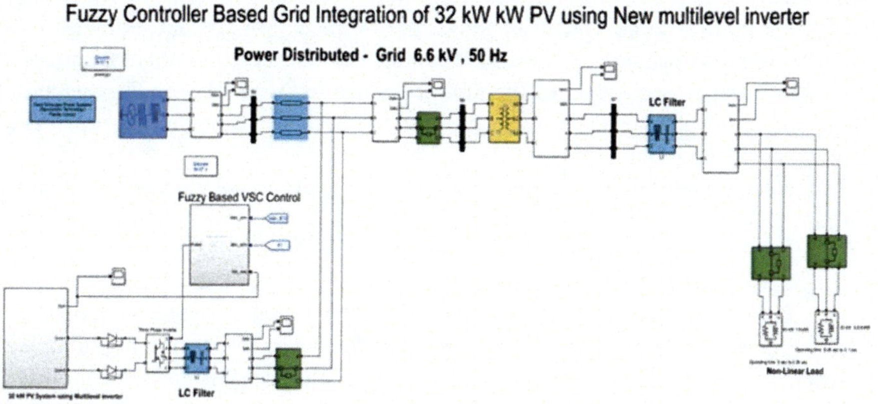

Fig. 14.16 MATLAB/Simulink model for proposed AI-based PV system

(VSC). Six PV sources are used in the proposed simulation model, and each has a power supply capacity of 5300 W [11]. The overall PV power is 32 kW interfaced to grid through new CMLI as shown in Fig. 14.15. In order to boost energy efficiency and lower the THD voltage value and current waveform, the new multilevel inverter has been connected to the microgrid system via an LC filter [6].

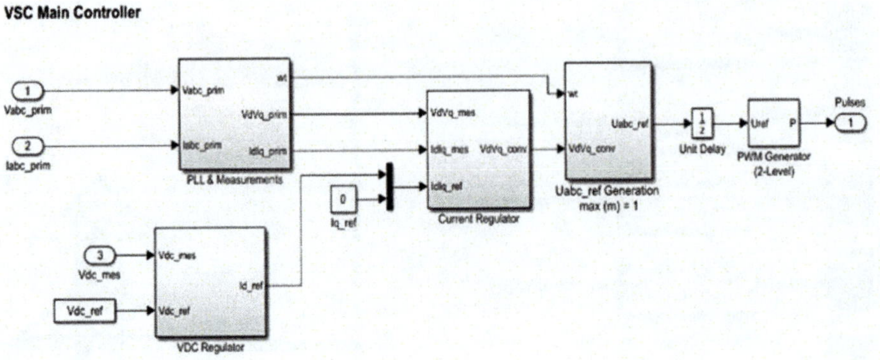

Fig. 14.17 VSC controller

14.4.2 Voltage Source Control Unit

The network integration mode, the voltage source-based controller, played an important role in PV and grid synchronization [12]. The voltage converter consists of three main subdivisions as seen in Fig. 14.17, such as a phase locked loop (PLL), voltage, and current control loop. The PLL allows the frequency of two AC sources to be matched. The grid voltage, current, and frequency are observed by this PLL loop, in which the voltage and current of the grid are converted from abc- to dq-axis and it is supplied to the new regulator for comparison with the performance parameters of the inverter.

The voltage control unit helps to control the output voltage of the inverter concerning the reference voltage. The voltage regulator matches the input voltage and reference voltage determined by the inverter. The PI controller is related to the existing current regulator by the error signal, and it produces a direct axis reference current. If the Id current is positive, the inverter produces the active power and the Iq current is positive, the inverter observes the reactive power, the current regulator will govern the inverter activity of the proposed device. The fuzzy controller for current regulators was developed in this article [6]. Two input signals and two output signals are on the fuzzy controller for, e.g., direct axis and quadrature axis current at the input side and direct axis and quadrature current regulated at the end of the output. The fuzzy membership function is built based on the fuzzification process of triangle method and the defuzzification process of centroid method [6].

14.4.3 Results and Discussion

The AI-based proposed configuration has been simulated and the total inverter (multilevel) power generating capacity is 32 kW. The suggested CMLI is aligned with the power grid and is seen in Fig. 14.18 with its current and voltage waveform.

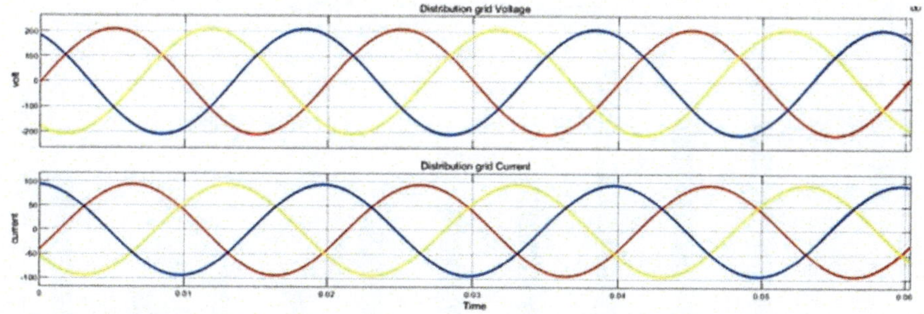

Fig. 14.18 Voltage and current waveform of the proposed system at the grid side

THD is a calculation that indicates how much distortion there is voltage or current is induced by the harmonics in the signal and also, it is a significant feature of the power grid. The proposed CMLI has a smaller number of switches and has 13-level output. If the number of levels at output side increases, THD gets reduced. The LC filter is often used in the middle of the grid and the CMLI terminal, reducing the excess THD in voltage and current waveforms. At which the new CMLI terminal less THD is achieved, which is evidence of the good power quality of the system. The proposed CMLI increased the quality of the PV power system connecting to the power grid by evaluating and addressing its current and voltage THD levels in which the THD voltage level is 0.10% and the current THD level is 0.15% as seen in Figs. 14.19 and 14.20, respectively. To show the efficacy of the method, the THD above was calculated in compliance with standard IEEE 519 and 1547.

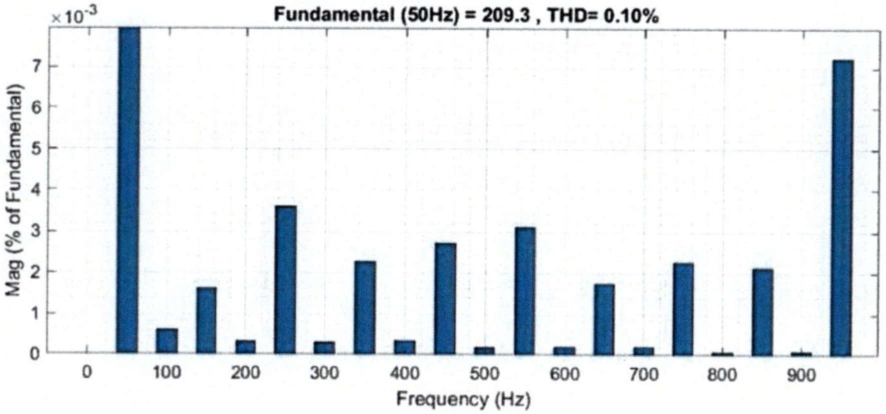

Fig. 14.19 THD for voltage waveform

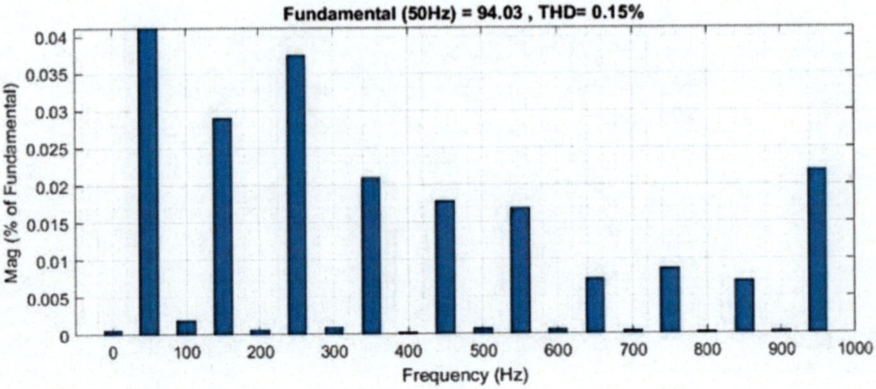

Fig. 14.20 THD for current waveform

14.5 Conclusion

In the MATLAB/Simulink software, the super-lift Luo three-stage converter was built in this paper and its output was evaluated under different operating conditions. The super lift—Luo converter based PV system is developed with fuzzy logic controller for tracking optimum point and it is simulated to analyze its performance under different weather conditions. In the MATLAB tool, the proposed PV-new CMLI grid integration was simulated. If the level of the new CMLI increased in the system, the complexity of circuit configuration will increase, as will the number of switches utilized in the system, which will increase the harmonic content of the PV system. The fuzzy logic controller was developed to provide multilevel inverter grid integration and analyzes its output in different operational circumstances and the effects of simulation with IEEE 519 & 1547 are analyzed to demonstrate proposed system quality. Finally, there is a noticeable increase in the energy efficiency for the grid-integrated PV system for the proposed multilevel inverter and controller and this proposed system is highly advocated for both grid and autonomous industrial applications for hybrid clean energy sources.

References

1. N.B. Deshmukh, R.D. Thombare, P.T. Jadhav, D.S. More, MISO Luo DC-DC converter for renewable energy applications, in *International Conference on Energy Systems and Applications 2015*, pp 63–68, Pune (2015)
2. J.R. Christina, J. Pradeep, Design and implementation of wind energy based improved positive output super-lift Luo converter, in *Second International Conference on Science Technology Engineering and Management (ICONSTEM-2016)*, pp. 287–292, Chennai (2016)
3. S.A. Nath, J. Pradeep, PV based design of improved positive output super-lift Luo converter. In: Second international conference on science technology engineering and management (ICONSTEM-2016), pp. 293–297, Chennai (2016)

4. G. Anandhakumar, M. Venkateshkumar, P. Shankar, Intelligent controller based MPPT method for the Photovoltaic power system, in *International Conference on Human-Computer Interactions (ICHCI-2013)*, pp. 1–6, Chennai (2013)
5. M. Venkateshkumar, R. Raghavan, N. Kumarappan, Design of a new multilevel inverter standalone hybrid PV/FC power system. Fuel Cells **15**, 862–875 (2015)
6. I. Mahendravarman, S.A. Elankurisil, M. Venkateshkumar et al., Artificial intelligent controller-based power quality improvement for microgrid integration of photovoltaic system using new cascade multilevel inverter. Soft. Comput. (2020). https://doi.org/10.1007/s00500-020-05120-2
7. F.A. Silva, Advanced DC/AC inverters: applications in renewable energy (F.L. Luo, H. Ye [Book News]). IEEE Ind. Electron. Magazine **7**(4), 68–69, Dec 2013
8. E. Babaei, M. Sarbanzadeh, M.A. Hosseinzadeh, C. Buccella, A new topology for cascaded multilevel inverters with a reduced number of power electronic switches, in *7th Power Electronics and Drive Systems Technologies Conference (PEDSTC-2016)*, pp. 165–170. Tehran (2016)
9. K. Chen et al., Cascaded iH6 multilevel inverter with leakage current reduction for transformerless grid-connected photovoltaic system, in *IEEE 12th International Conference on Power Electronics and Drive Systems (PEDS-2017)*, pp. 613–617. Honolulu, HI (2017)
10. M. Venkateshkumar, R. Raghavan, Fuel cell power penetration into AC distribution grid by using new cascade multilevel inverter with minimum number of switches, in *International Conference on Smart Technologies and Management for Computing, Communication, Controls, Energy and Materials (ICSTM-2015)*, pp. 568–574. Chennai (2015)
11. M. Venkateshkumar, AI based Hybrid PV-FC and Wind Power system using MLI (2020). https://www.mathworks.com/matlabcentral/fileexchange/69477-ai-based-hybrid-pv-fc-and-wind-power-system-using-mli. MATLAB Central File Exchange. Retrieved 31 July 2020
12. E. Babaei, M.A. Hosseinzadeh, M. Sarbanzadeh, C. Cecati, A new basic unit for cascaded multilevel inverters with a reduced number of power electronic devices, in *7th Power Electronics and Drive Systems Technologies Conference (PEDSTC-2016)*, pp. 185–190. Tehran (2016)

Chapter 15
Power Quality Enhancement in Single Stage Non-inverted Output Bridgeless Buck–Boost Converter

J. Gnanavadivel, S. Muralidharan, and S. Joe Magellah

Abstract This research manuscript proposes a bridgeless single phase buck–boost power factor correction converter providing non-inverted DC load voltage. Proposed topology has two power switches in the current path over every switching period, and moreover, this topology does not have bridge rectifier at the input side. Therefore, it leads to improved thermal management and reduced conduction losses. This further increases the efficiency of the system. The suggested converter offers unity input as it is operated in discontinuous conduction mode. Harmonic content in supply current is greatly reduced improving the power quality indices to the recommended standard. In this manuscript, performance analysis of the proposed converter is observed, and its implementation is compared with standard buck–boost converter.

Keywords Power factor correction (PFC) · Discontinuous conduction mode · (DCM) · Bridgeless buck–boost converter · Total harmonic distortion (THD)

15.1 Introduction

In recent years, in order to stoke up the power factor, in AC-DC conversion, the roles of power factor correction converters are predominant. Harmonic present in line current is reduced greatly [1]. With the intention to enhance the efficiency of the system or a transmission line, the power factor at the load side has to be maintained near unity. If power factor gets diminished outside certain limits, then the consumer has to pay additional cost as the active power utilized by the consumer will get increased.

In the conventional AC-DC converter, the presence of diode bridge rectifier (DBR) induces a high conduction loss which in turn reduces the system efficiency. Therefore, we go for bridgeless PFC converter with more modest quantity of active switches operated during every switching period. Also, the abolition of input DBR in bridgeless topology trim down the conduction losses, size and cost associated

J. Gnanavadivel (✉) · S. Muralidharan · S. Joe Magellah
EEE Department, Mepco Schlenk Engineering College, Sivakasi, Tamil Nadu 626005, India
e-mail: smurali@mepcoeng.ac.in

R. Seyezhai et al. (eds.), *Recent Trends in Renewable Energy Sources and Power Conversion*, Springer Proceedings in Energy,
https://doi.org/10.1007/978-981-16-0669-4_15

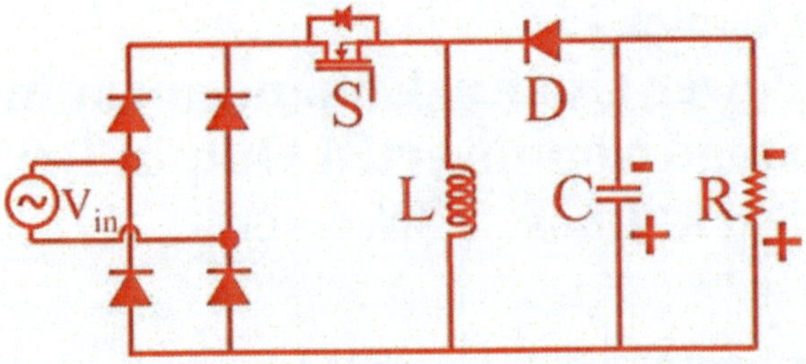

Fig. 15.1 Buck–boost converter of conventional nature

with the system. In this context, various AC-DC power converters-based bridgeless configuration have been analysed in [2].

For power factor correction, the conventional boost [3] and buck converter are the most popular front-end converters. However, they have issues like inadequate feature in current shaping and cycle restrictions. The PFC converter at the power conversion always offers a high voltage in output side comparative to the entry side and does not give wide range of input voltage regulation, deprived load efficiency while operating in universal input voltage. This creates challenge while operating boost converter as universal PFC converter [4, 5]. The buck PFC converter can step down the voltage to support small output voltage applications, but they cannot bear the input current during the course of the cycle and highly distort the line current. But the buck–boost PFC bridgeless converter does both the buck and boost operation offering a wide span of input voltage regulation, low distortion of line current with high efficiency, improved power factor and reliable operation. There are various topologies of buck–boost converter available for PFC applications. Buck–boost converter of conventional nature is shown in Fig. 15.1 with the presence of DBR.

15.2 Proposed System

A new single stage buck–boost PFC converter offering positive output voltage is introduced in this paper. The suggested converter is made to operate discontinuous conduction mode (DCM). Usually, the PFC converters are considered to function in (CCM)/(DCM) [6]. Sensing of line voltage/current and DC link voltage is needed in the PFC operating in CCM mode for PFC operation. However, PFC converter in DCM operation needs one sensor needed towards DC link voltage control, and on higher stresses on the switch, the PFC mode operation is realized at the AC mains. Thus, converter for PFC functioning in DCM is chosen for reduced power requirement applications. Venkatesh et al. [6] brief about a novel AC-DC converter with discontinuous conduction mode.

In [6, 7], the PFC converters of buck–boost operation are employed which have output voltage of negative nature. Therefore, an inverting amplifier circuitry which is meant to converter negative voltage to non-inverted voltage is additionally needed [8]. This requirement of extra inverting amplifier circuit increases the cost. In [8, 9], modified AC-DC sepic converter for power quality correction has been implemented with boost operation only.

In contrary, the authors have implemented a unique buck–boost converter. This manuscript introduces a buck–boost converter of bridgeless nature for PFC applications with positive output voltage, high efficiency, less line current distortion and high power density. DCM operating mode is chosen for this proposed converter topology. Figure 15.2 shows the circuit configuration of single stage buck–boost converter for PFC with inverted output voltage.

To achieve the output voltage of positive nature exclusive of inverse amplifier circuit, polarity of devices involved in the circuit needs to be reassigned in Fig. 15.2, so that we get the proposed bridgeless operation buck–boost converter for PFC as provided in Fig. 15.3. This makes the feedback control circuit simpler with reduced cost compared with buck–boost PFC converter having negative side output voltage requiring an inverse amplifier circuit for feedback control. The proposed PFC-based

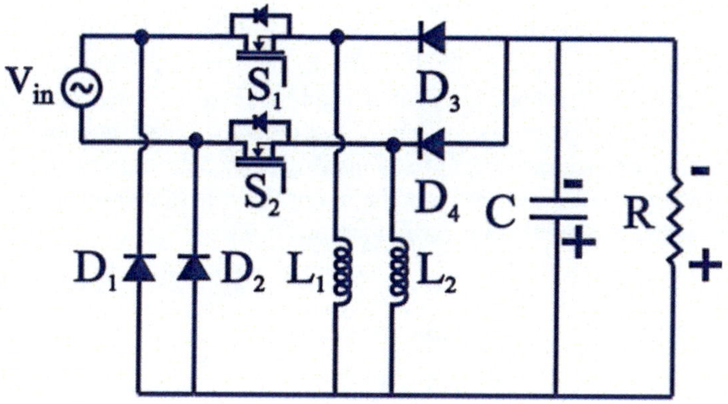

Fig. 15.2 Single stage inverted buck–boost PFC converter

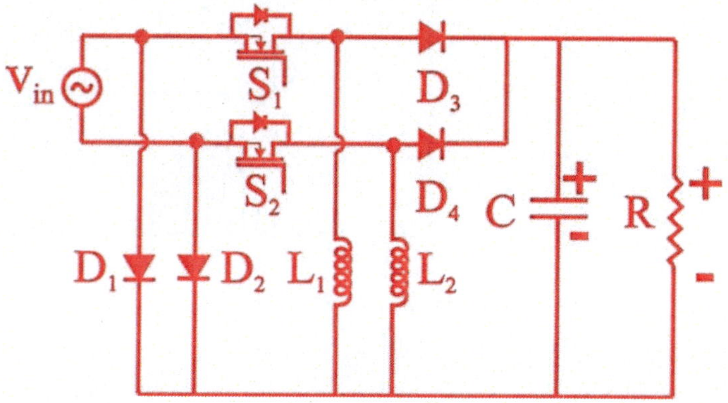

Fig. 15.3 Single stage non-inverted buck–boost PFC converter

single stage buck–boost converter in Fig. 15.3 is made up of diodes ($D3$, $D4$), switches ($S1$, $S2$) and two inductors ($L1$, $L2$), which are recognized as two buck–boost cells.

15.3 Proposed Single Stage Non-inverted Buck–Boost Converter Providing PFC Operation

Performance of the proposed PFC converter is rendered into two sections. One is input voltage line cycle operation, and another is operation during complete switching cycle.

15.3.1 Input Voltage Line Cycle Operation

In upper half cycle of AC, the S_2 switch is active, and switch S_1 is idle. In other half cycle of AC waveform, the S_1 switch is active, and switch S_2 is inactive. Thus, in positive cycle of AC, the diode D_1, switch S_2 and inductor L_2 are in conduction, and in negative cycle, the diode D_2, switch S_1 and inductor L_1 are in conduction. This proposed converter has three ways of action in AC. Discontinuous nature of inductor current in the proposed converter is shown in Fig. 15.4. In the waveform, the current in inductor is discontinuous for a certain period.

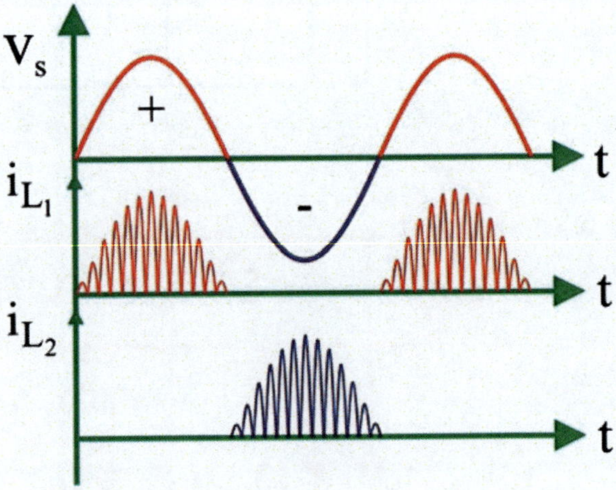

Fig. 15.4 Waveform of supply voltage during upper and lower half cycles

15.3.2 Operation Over a Complete Switching Cycle

The operating cycle of the PFC converter presented has three modes during positive half cycle as well as during negative half cycle.

Mode I: In this Mode I, the diode D_1 gets forward biased and charges the inductor L_2. Then, the current in inductor (i_{L2}) rises. Then, switch S_2 gets closed so that DC link capacitor C_1 discharges to load as presented in Fig. 15.5.

Mode II: Here, stored energy of inductor L_2 gets transferred to C_1, as switch S_2 remains off conditions. This process continues until inductor L_2 completely discharges its stored energy, and so the current via L_2 diminishes and goes to zero as given in Fig. 15.5.

Mode III: Here, current through L_2 turns into zero for the leftover switching interval as no inductor energy is left out. Here, none of the components are conducting except for output side DC link capacitor C_1 which is feeding the DC load. This operating sequence is repetitive whenever S_2 is turned during completion of every switching cycle.

Likewise the identical operation happens in negative cycle of the voltage in supply side with the switch S_1, diode D_2 and inductor L_1 in the conduction state.

15.4 Design of Bridgeless Buck–Boost PFC Converter

The proposed single stage buck–boost PFC converter providing non-inverted DC load voltage is designed to function in DCM in such a way that inductors current becomes discontinuous in switching phase. The bridgeless converter presented is intended for a DC link voltage of 50 V and the power rating of 100 W. For a supply

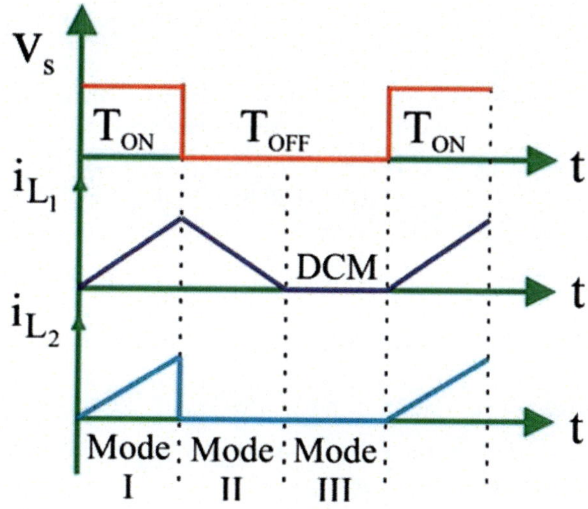

Fig. 15.5 Complete switching cycle waveforms

Table 15.1 Values of parameter for the proposed PFC converter

Parameters	Values
Line voltage (V_{in})	230 V
Output voltage (V_{dc})	50 V
Frequency of switching (f_s)	20 kHz
Duty cycle (D)	18%
Source frequency (f)	50 Hz
Inductor values (L_1, L_2)	50 μH
DC link capacitor ($C1$)	4000 μF
Resistance	23.04

voltage of 230 V, buck–boost converter duty cycle (D) is given as [5], in which V_{dc}—Load voltage and V_{in}—RMS input voltage.

$$D = \frac{V_{dc}}{V_{dc} + V_{in}} \tag{15.1}$$

in which, V_{dc}—Load voltage and V_{in}—RMS input voltage.

The value of input inductors L_1, L_2 of suggested converter that operates in critical mode of conduction is given as [8]

$$L_1, L_2 = \frac{R(1 - D)^2}{2f_s} \tag{15.2}$$

where D—duty ratio, R—equivalent resistance in load, representing the duty ratio and f_s—switching frequency.

The DC link capacitor C_1 is given as [8]

$$C_1 = \frac{1}{2\omega\Delta V_{dc}} \tag{15.3}$$

in which ΔV_{dc} is the ripple voltage in DC link. Value of DC link capacitor with permitted ripple of 3% is numerically calculated for obtaining desired values of DC link voltage.

The design parameters required for the converter to manage the DC link voltage are shown in Table 15.1.

15.5 Simulation Results

With an objective to validate the implementation of the proposed system, a comparative simulation study between the conventional and the proposed system is conducted.

15.5.1 Control of Converter

In this system, the control parameter is the voltage. Here, a proportional integral (PI) controller is used for this purpose. The reference voltage (V_{dc}^{*}) is given as

$$V_{dc}^{*} = K_{v} V_{ref}^{*} \tag{15.4}$$

The error voltage generated from the comparator is given as

$$V_{e}(n) = V_{dc}^{*}(n) - V_{dc}(n) \tag{15.5}$$

where n stands for nth sample.

To generate regulated PFC output (VCC) using the error signal (Ve), the expression is given as follows

$$V_{cc}(n) = V_{cc}(n-1) + K\{V_{e}(n) - V_{e}(n-1)\} + K_{i} V_{e}(n) \tag{15.6}$$

in which K_{p} and K_{i} are controller gains of proportional and integral part.

By using Ziegler Nichols method of tuning, the Kp and K_{i} values of the controller are obtained as 0.0001 and 0.25. Implementation of this PI controller helps in getting regulated output voltage from the converter.

15.5.2 Results

The converter topology is simulated using MATLAB implementing the design parameter. Figures 15.6 and 15.7 depict the waveforms obtained from simulation results pertaining to the planned and the conventional converter supplying for 100 W load with a 230 V AC as input. Figures 15.6a and15.7a show the waveforms of supply side current/voltage of proposed power converter and buck–boost PFC converter of conventional nature. Figures 15.6b and 15.7 show the total harmonic distortion of the projected PFC converter and conventional buck–boost PFC converter. Hereby, it is inferred that the planned PFC converter has harmonic distortion of about 1.9% which is less compared to that of the conventional converter which has harmonic distortion about 5.55%.

Graph depicting the effectiveness of the proposed PFC converter for various load power against classic configuration is shown in Fig. 15.8.

The conduction losses are high in diode bridge rectifier-based PFC converter. As the proposed topology does not have any diode bridge, the conduction losses have come done. This results in improved efficiency of the proposed converter. The high value of efficiency for a given load in Fig. 15.8 clearly shows the reduction in losses.

Table 15.2 compares the source current THD of conventional and proposed buck–boost converter with Refs. [10, 11]. It clearly favours the proposed converter as the

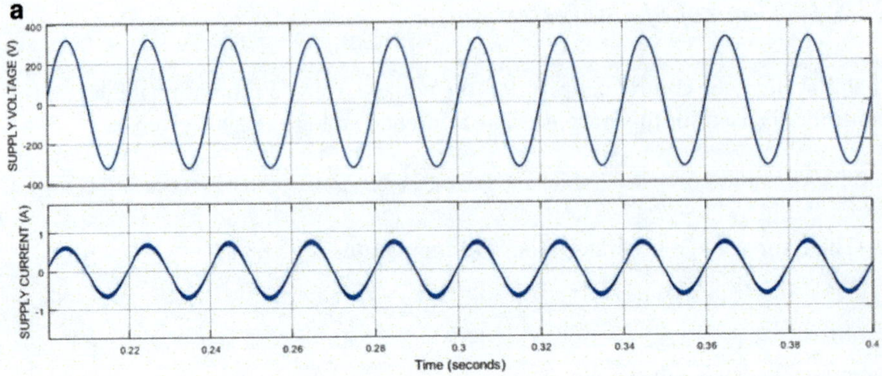

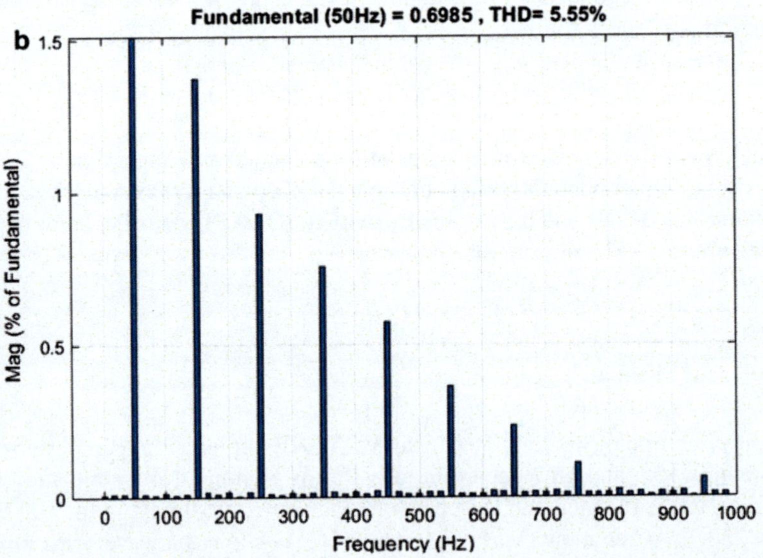

Fig. 15.6 **a** Simulated waveforms of supply voltage/current for buck–boost PFC conventional converter. **b** FFT analysis of total harmonic distortion of the conventional buck–boost PFC converter

THD level is very low compared to others. This reduction in THD level addresses the power quality issues cited in [12–18].

15.6 Conclusion

In this research paper, an innovative topology belonging to buck–boost type bridgeless configuration of PFC converter is proposed and discussed. It is inferred from the implementation of the proposed topology, and due to lower value conduction

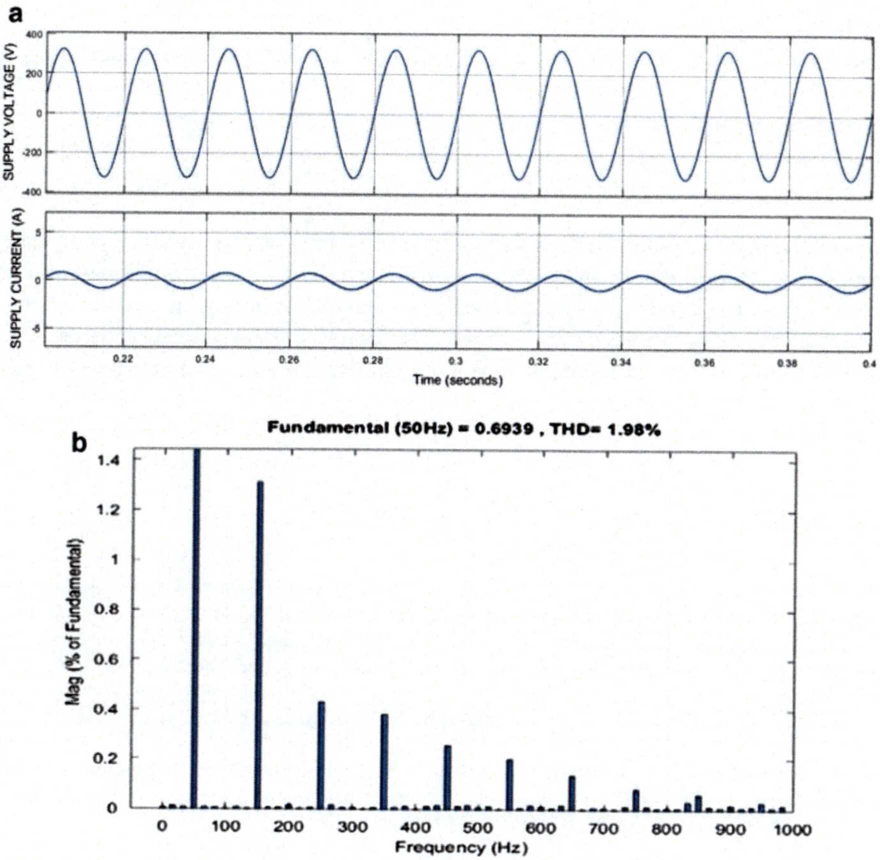

Fig. 15.7 **a** Simulated waveforms of line voltage/ current of the proposed converter. **b** FFT analysis of total harmonic distortion of the proposed bridgeless buck–boost PFC converter

Fig. 15.8 Graph showing the comparison of proposed and the conventional buck–boost PFC converter

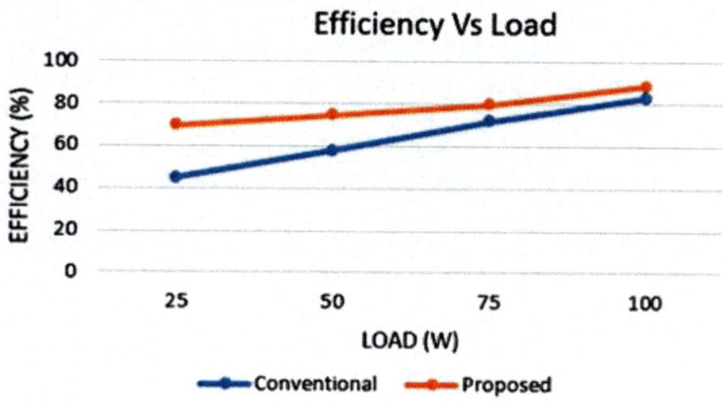

Table 15.2 Performance comparison

Parameter	Conventional buck–boost converter	Ref. [10]	Ref. [11]	Proposed buck–boost converter
Source current THD	5.55%	3.39%	14%	1.98%
Input Power Factor	0.98	1	0.98	1

as well as switching losses, there is an increase in the conversion efficiency while comparing with the classical type buck–boost converter for power factor correction. This proposed topology gives positive output voltage eliminating the need of inverse amplifier circuit at the feedback control. The THD is also reduced to below 2%. Thus, the proposed system proves to be more efficient compared to the conventional converters.

References

1. C. Qiao, K. Smedley, A topology survey of single-stage power factor corrector with a boost type input-current-shaper. IEEE Trans. Power Electron. **16**(3), 360–368 (2001)
2. B. Singh, B.N. Singh, A. Chandra, K. Al-Haddad, A. Pandey, D.P. Kothari, A review of single-phase improved power quality AC-DC converters. IEEE Trans. Ind. Electron. **50**(5), 962–981 (2003)
3. Y. Jang, M.M. Jovanovic, Bridgeless high-power factor buck converter. IEEE Trans. Power Electron. **26**(2), 602–611 (2011)
4. J. Gnanavadivel, P. Yogalakshmi, N.S. Kumar S. Muralidhran, High efficient single stage Cuk LED driver for universal input voltage applications with improved power quality. Optoelectron. Adv. Mater. Rapid Commun. **12**(11–12), 694–699 (2018)
5. A.J. Sabzali, E.H. Ismail, M.A. Al-Saffar, A.A. Fardoun, New bridgeless DCM Sepic and Cuk PFC rectifiers with low conduction and switching losses. IEEE Trans. Indus. App. **47**(2), 1179–1188 (2011)
6. V. Bist, B. Singh, An adjustable-speed PFC bridgeless Buck–Boost converter-fed BLDC motor drive. IEEE Trans. Ind. Electron. **61**(6), 2665–2677 (2014)
7. J. Gnanavadivel, P. Yogalakshmi, N. Senthil Kumar, K. S. Krishna Veni, Design and development of single phase AC–DC discontinuous conduction mode modified bridgeless positive output Luo converter for power quality improvement. IET Power Electron. 12(11), 2722–2730 (2019)
8. J. Gnanavadivel, N. Senthil Kumar, S.T. Jaya Christa, S. Muralidharan, Design and implementation of FPGA-based high power LED lighting system for ships. Indian J. Marine Sci. **48**(5), 724–729 (2019)
9. S. Muralidharan, S. Menaka, Application of evolutionary algorithms for harmonic profile optimization in symmetric multilevel inverter used in medical electronic equipments. Curr. Signal Transduct. Therapy **14**(1), 12–20 (2019)
10. M. Junaid, B. Singh, Analysis and design of buck-boost converter for power quality improvement in high frequency on/off-line UPS system, in *2014 IEEE International Conference on Power Electronics, Drives and Energy Systems (PEDES), Mumbai*, pp. 1–7 (2014)
11. R. Lin, R. Wang, Non-inverting buck-boost power-factor-correction converter with wide input-voltage-range applications, in *IECON 2010—36th Annual Conference on IEEE Industrial Electronics Society, Glendale, AZ*, pp. 599–604 (2010)
12. H. Esam, Ismail: bridgeless SEPIC rectifier with unity power factor and reduced conduction losses. IEEE Trans. Ind. Electron. **56**(4), 1147–1157 (2009)

13. B. Zhao, A. ABranovitz, K. Smedley, Family of bridgeless Buckboost PFC rectifiers. IEEE Trans. Power. Electron. **30**(2), 6524–6527 (2015)
14. J. Chen, D. Maksimovic, R.W. Erickson, Analysis and design of a low-stress buck-boost converter in universal-input PFC applications. IEEE Trans. Power Electron. **21**(2), 320–329 (2006)
15. S. Muralidharan, S. Menaka, Novel symmetric and asymmetric multilevel inverter topologies with minimum number of switches for high voltage of electric ship propulsion system. Indian J. Geo Marine Sci. **46**(09), 1920–1930 (2017)
16. Saranya Paramashivam, S. Muralidharan, Analysis of AC/AC converter with enhanced power quality, in *3rd IEEE International Conference on Electronics Computer Technology (ICECT)*, vol. 2, pp. 179–183 (2011)
17. S. Muralidharan, S. Menaka, Design and performance analysis of novel boost DC-AC converter, in *3rd IEEE International Conference on Electronics Computer Technology (ICECT)*, vol. 2, pp. 168–172 (2011)
18. L. Huber, Y. Jang, M.M. Jovanovic, Performance evaluation of bridgeless PFC boost rectifiers. IEEE Trans. Power Electron. **23**(3), 1381–1390 (2008)

Chapter 16
Comparative Study for the Estimation of the Rays of Sun Incident on Inclined Plane and Flat Plane by Isotropic and Anisotropic Sky Models for Vadodara, Gujarat

Sriparna Das and Kumari Namrata

Abstract The requirement to learn is to make comparison of various models for predicting the amount of the rays from the sun coming on an inclined plane and on the horizontal plane. To make this comparison, various models were used some of which are isotropic while others are anisotropic to calculate the total amount of rays in a month on Vadodara, Gujarat, situated in the western part of India. The inclined angle was assumed to be 22 °N of Vadodara. The results getting from various formulas were compared to the actual data. It was seen that Hay and Davies model (H.D.M) shows the maximum part of solar rays' incident on an inclined plane throughout the year, whereas Badescu model (B.A.M) estimated the lowest portion of solar rays of sun among all the isotropic and anisotropic models. Finally, it is shown that BAM was preferred than other models due to less statistical error, and the predicted solar radiation is very near to the measured data.

Keywords Solar radiation · Radiation calculation · Inclined surface · Various models · Gujarat

16.1 Introduction

Solar radiation is the sum of portion of solar rays emitted by sun on the Earth's surface. Nowadays, the world is moving into the use of renewables. In case of grid system, there is a problem in supplying the energy to remote areas due to high loses of transmission line, and also it is difficult to place transmission line all the areas [1]. So, the world is now moving into the concept of micro-grid. Micro-grid is basically the grid that connects small areas. So basically, DGs that are distributed generations are placed near the distribution end which supply to the load of small area. So, if there

S. Das (✉) · K. Namrata
Department of Electrical Engineering, NIT Jamshedpur, Jamshedpur, Jharkhand 831014, India

K. Namrata
e-mail: namrata.ee@nitjsr.ac.in

R. Seyezhai et al. (eds.), *Recent Trends in Renewable Energy Sources and Power Conversion*, Springer Proceedings in Energy,
https://doi.org/10.1007/978-981-16-0669-4_16

is fault in the utility grid, the area where DGs are connected, the supply still remains over there [2]. So renewable energy, like solar and wind, is implemented in it. So, it is important to study the area where the amount of solar radiation is maximum. It helps in increasing the efficiency of producing the energy. So, prediction of solar rays falling on a particular area is important before establishing one solar plant.

The total solar radiation falling on flat surface are the components of diffuse and direct radiation so, finding individual of them is necessary in case of solar radiation calculation. There is no such availability of measured data for solar rays falling on the inclined surface. The solar rays falling on the inclined surface can be calculated from changing the solar rays on a flat surface to that falling on the inclined surface for longer time.

Solar rays falling on any surface are basically the sum of three parts: direct radiation, diffuse radiation and reflected radiation from surface [3]. The main theme of this work is to calculate mean global, diffused and beam radiation for a month and a day on the flat plane in Vadodara, Gujarat, taking the help of six formulas and finding of total solar rays falling at inclined plane of inclination angle 22 °N (considered to be latitude for place Vadodara) taking help of the six formulas and lastly to compare each formulas result with the actual information taking help of methods such as MAPE, MBE, *t*-stat and RMSE [4].

16.2 Methodology and Study of Locations

16.2.1 Study of Location

The geographical location of Vadodara city in Western India North latitude 22 °N and East longitude is 73° 10′ and altitude elevation of 39 m or 128 ft. The city lies on the bank of Vishwamitri River in Central Gujarat. The Vishwamitri River in summer dries up and becomes small steam of water. The city has roughly 800 mm precipitation and has semiarid climate. There are basically three timings of a year: scorching hot, rainy and winter. Besides from monsoon season, the climate is basically dry. The weather from March to July is hot and has an average maximum temperature of 45 °C and average minimum temperature of 23 °C. The average max temperature from November to February is 29–31 °C, and min temperature is from 14 to 16 °C. The wind responsible for mild cold in time of December to February is northerly wind, and effect of south-west monsoon makes climate humid during the time of mid-June to mid-September. The average rainfall is 93 cm (37 inches).

16.2.2 Solar Rays on the Flat Plane with No Inclination

H_{01} on the flat plane with no tilt of surface is given by the expression [5]:

$$H_{01} = \frac{24 I_{SC1}\left(1 + 0.033 \cos \frac{360 N_1}{365}\right)(\omega_{S1} \sin \varphi_1 \sin \delta_1 + \cos \varphi_1 \cos \delta_1 \sin \omega_{S1})}{\pi \text{ KWh/m}^2 \text{ day}} \tag{16.1}$$

I_{SC1} has value of 1.367 kW/m^2, N_1 is counted in number of days of the years, and ω_{S1} and δ_1 are measured in degrees. The declination angle (δ_1) mathematically is given by Cooper's (1969) equation:

$$\delta_1 = 23.34 \sin \frac{360}{365}(284 + N_1) \tag{16.2}$$

where N_1 is the day counting January 1, and the ω_{S1} is related to angle delta and phi as they are expressed in the formula represented by

$$\omega_{S1} = \cos^{-1}(-\tan \delta_1 \tan \phi_1) \tag{16.3}$$

H_{G1} which is formulated by Angstrom is shown by the below equation:

$$\frac{H_{G1}}{H_{01}} = a_1 + b_1\left(\frac{s_1}{s_{\max 1}}\right) \tag{16.4}$$

s_1 is considered to be length of solar day calculated in hours. For Vadodara, the value of average length of solar day is 9.5 h. $s_{\max 1}$ is the length of the longest solar day in the month (hour). The formula of $s_{\max 1}$ is given by

$$s_{\max 1} = \left(\frac{2}{15}\omega_{S1}\right) \tag{16.5}$$

a_1 and b_1 are the constants determined for various cities in the world by measurement. The value of $a = 0.28$ and $b = 0.48$ is for Vadodara. The formulas for calculating the Angstrom's constants are given by

$$a_1 = 0.409 + 0.5016 \sin(\omega_{S1} - 60) \tag{16.6}$$

$$b_1 = 0.6609 + 0.4767 \sin(\omega_{S1} - 60) \tag{16.7}$$

The value of a_1 and b_1 for Vadodara si given by M.S in 1979. G.G in 1985 proposed the equation for calculating the diffused radiation that is

$$\frac{H_{D1}}{H_{G1}} = 0.8677 - 0.7365\left(\frac{s_1}{s_{\max 1}}\right) \tag{16.8}$$

Then monthly average beam radiation is given by

$$H_{B1} = H_{G1} - H_{D1} \tag{16.9}$$

16.2.3 Solar Rays Falling on the Tilted or Inclined Plane

Solar rays falling on inclined surface are the sum of direct ray from sun, ray getting passed through air by deflecting in atmosphere particles and the ray coming by striking from the ground [6]. It is denoted by (H_{T1}) can be expressed as follows:

$$H_{T1} = H_{TB} + H_{TR} + H_{TD} \tag{16.10}$$

Direct Radiation (H_{TB}): Direct radiation on the inclined plane is shown below:

$$H_{TB} = H_{B1} R_{B1} \tag{16.11}$$

R_{B1} is known as tilt factor given by

$$\begin{aligned} R_{B1} &= \frac{\cos}{\cos_{Z1}} \\ &= \frac{\sin \delta_1 \sin(\varphi_1 - \beta_1) + \cos \delta_1 \cos(\varphi_1 - \beta_1) \cos \omega_1}{\sin \delta_1 \sin(\varphi_1) + \cos \delta_1 \cos(\varphi_1) \cos \omega_1} \end{aligned} \tag{16.12}$$

All the parameters in the above equation are measured in degree.

Reflected Radiation (H_{TR}): Reflected radiation is basically the component we get when solar rays are striking from surface and remaining surroundings objects. It is given by

$$R_{R1} = (1 - \cos \beta_1)\frac{\rho_1}{2} \tag{16.13}$$

β_1 is measured in degree, and ρ_1 is dependent on the place where the calculations need to be done. It varies from 0.2 to 0.9 depending upon the temperature.

Disperse Radiation (H_{TD}): Disperse radiation is the ratio of flux of diffuse ray of sun falling on the inclined plane to that on the flat surface with inclination zero [7]. It is given

$$R_{D1} = \frac{(1 + \cos \beta_1)}{2} \tag{16.14}$$

β_1 is measured in degree.

16.2.4 Brief Explanation of Isotropic and Anisotropic Empirical Formulas

There are various formulas to find the amount of radiation scattered by displacing in atmosphere on the inclined surface [8]. They are divided as isotropic and anisotropic formula models. Here, in this paper, basically some formulas are selected and the total incident solar ray from sun on the inclined surface is formulated by taking help of these formulas to find the maximum radiation on Vadodara, Gujarat. The three isotropic formulas [9] are named as L.J.M (Liu & Jordan), K.O.M (Koronakis) and B.A.M, and anisotropic formulas are named as H.D.M, R.E.M (Reindl et al.) and H.D.K.R (Hay and Davies, Klucher and Reindl model) [10] were analyzed. The explanation of the formulas for the tilted surface is discussed as below:

L.J.M: This formula basically helps to find the total amount of solar rays on the inclined plane consisted of three main parts. They are beam radiation, radiation hitting back from ground and radiation deflected in atmosphere due to scattering. The overall formula is calculated by considering these three radiations on an inclined surface. This isotropic model formula is given as follows:

$$H_{T1} = H_{B1}R_{B1} + \frac{H_{G1}(1 - \cos\beta_1)\rho_1}{2} + \frac{H_{D1}(1 + \cos\beta_1)}{2} \tag{16.15}$$

K.O.M: In this model, it is considered for isotropic sky radiation that the slope is 90° which will give 66.7% of total solar ray deflected in atmosphere or is scattered. He gave the below formula to calculate the radiation of the tilted surface.

$$H_{T1} = H_{B1}R_{B1} + \frac{H_{G1}(1 - \cos\beta_1)\rho_1}{2} + \frac{H_{D1}(2 + \cos\beta_1)}{3} \tag{16.16}$$

B.A.M: This has given the expression in measuring total amount of solar ray on the inclined plane which is given as follows:

$$H_{T1} = H_{B1}R_{B1} + \frac{H_{G1}(1 - \cos\beta_1)\rho_1}{2} + \frac{H_{D1}(3 + \cos 2\beta_1)}{4} \tag{16.17}$$

H.D.M: H.D.M tells that the ray of the sun that is hitting and coming back from sky consists of two components basically isotropic and other is circumsolar. It is predicted that the ray coming from sun is circumsolar and coming from other part of sky is isotopical which consists of diffused part, and these are measured according to anisotropy index (A_1). The estimated ray falling on the tilted plane is given by formulas as follows:

$$\begin{aligned} H_{T1} = {} & (H_{B1+}H_{D1}A)R_{B1} + H_{G1}(1 - \cos\beta_1)\frac{\rho_1}{2} \\ & + H_{D1}\frac{(1 + \cos\beta_1)}{2}(1 - A_1) + A_1 R_{B1} \end{aligned} \tag{16.18}$$

where A_1 is basically anisotropy index, function of direct rays and extraterrestrial rays which is defined by

$$A_1 = \frac{H_{B1}}{H_{01}} \tag{16.19}$$

R.E.M: Basically, comparing to the other models, one thing that is added to the H.D.M is the amount of brightness part in the formula. A_1 was established in the formula of H.D.M, $f_1 = \sqrt{\frac{H_{B1}}{H_{G1}}}$ was added to the term with $\sin^3\left(\frac{\beta}{2}\right)$ for horizontal brightness, and final solar rays on inclined plane are shown as follows:

$$H_{T1} = (H_{B1+}H_{D1}A)R_{B1} + H_{G1}(1\cos\beta_1)\frac{\rho_1}{2} + H_{D1}\left\{\frac{(1+\cos\beta_1)}{2}(1-A_1)\left[1+\sqrt{\frac{H_{B1}}{H_{G1}}}\sin^3\left(\frac{\beta}{2}\right)\right] + A_1 R_{B1}\right\} \quad (16.20)$$

H.D.K.R.M: He considered the sum of beam radiation along with rays named isotropic, horizontal brightening and circumsolar and developed H.D.K.R model which is expressed as follows [11]:

$$H_{T1} = (H_{B1+}H_{D1}A)R_{B1} + H_{G1}(1\cos\beta_1)\frac{\rho_1}{2} + H_{D1}\left\{\frac{(1+\cos\beta_1)}{2}(1-A_1)\left[1+\sin^3\left(\frac{\beta}{2}\right)\right]\right\} \quad (16.21)$$

16.2.5 Methods of Evaluation of Models

In this paper, the mean hourly diffuse and global radiation for Vadodara are calculated and are compared with measured data by the Indian Meteorological Department, and for an inclined surface the total radiation is calculated [12]. The judgment of which formula is better is done by performing the following method:

a. M.A.P.E_1
b. M.B.E_1
c. R.M.S.E_1
d. T-stat

M.A.P.E_1: M.A.P.E_1 helps to calculate how correct is the input set of data in the form of percent and is given by

$$\text{M.A.P.E}_1 = \frac{1}{n_i}\sum_{i=1}^{n_i}\frac{(I - I_{P)}}{I} * 100 \quad (16.22)$$

In case of perfectly accurate surface, the value of M.A.P.E_1 is neither positive nor negative number.

M.B.E_1: M.B.E_1 gives long-term performance as it helps to compare how close or what is the change of the formula which used data with the original data of a place.

It is basically the average deviation of predicted data from original data shown as follows:

$$\text{M.B.E}_1 = \frac{1}{n_i}\sum_{i=1}^{n}(I_{Pi} - I_i) \quad (16.23)$$

where I_i denotes ith non-imaginary, I_{Pi} denotes ith formulated data, and n denotes observations. Low M.B.E$_1$ is desired. In ideal case, null M.B.E$_1$ is best.

R.M.S.E$_1$: This provides small time process and is basically change formulated with actual data which is given by

$$\text{R.M.S.E}_1 = \sqrt{\frac{1}{n_i}\sum\nolimits_{i=1}^{n}(I_{Pi} - I_i)^2} \quad (16.24)$$

where I_i denotes ith non-imaginary part, I_{Pi} is the ith predicted part, and n denotes data count. The R.M.S.E$_1$ has non-negative value, and best is null (Table 16.1).

T-statistic (*t*-stat): M.B.E$_1$ and R.M.S.E$_1$ do not give accurately which model is to be selected and is best so it might create wrong choosing of accurate formulas [14]. It is possible to have large R.M.S.E$_1$ value and low M.B.E$_1$ value and vice versa. *T*-stat is given by

$$t\text{-stat} = \sqrt{\frac{(n_i - 1)\text{M.B.E}_1^2}{\text{R.M.S.E}_1^2 - \text{M.B.E}_1^2}} \quad (16.25)$$

Table 16.1 Predicted and measured H_{mt} and H_t in KWh/m^2 day falling on inclined plane taking help of anisotropic formulas [13]

Month	Measured (H_{mt})	L.J	K. O	B. A	H. D	R. E	H. D. K. R
January	5.82	6.14	6.15	6.09	6.33	6.66	6.34
February	6.33	6.42	6.44	6.37	6.59	6.99	6.59
March	8.01	6.70	6.72	6.63	6.81	7.28	6.82
April	6.93	6.73	6.75	6.64	6.77	7.27	6.77
May	8.15	6.58	6.60	6.49	6.57	7.08	6.57
June	6.06	6.44	6.47	6.35	6.40	6.91	6.41
July	5.68	6.46	6.49	6.37	6.43	6.93	6.44
August	5.29	6.65	6.67	6.56	6.67	7.18	6.68
September	5.54	6.69	6.71	6.62	6.78	7.26	6.79
October	6.70	6.49	6.50	6.43	6.64	7.06	6.64
November	4.72	6.15	6.16	6.10	6.33	6.69	6.34
December	4.86	6.00	6.01	5.96	6.20	6.52	6.21

16.3 Graphs for Comparison

The values of various models are plotted to make a detailed comparison among all the methods, and further conclusion can be made about their performances by determining the error calculations.

16.4 Results and Discussion

The taken variables for evaluation of mean solar rays of sun are calculated by using the above formulas for every month. Here calculation for decline angle is given according to Cooper's formula. The hour angle (ω_S) is calculated for different declination angle as the number of days changes for each month calculation. The monthly average daily sunshine hour is considered to be 9.5 h for the place of Vadodara, Gujarat. Using the formula which is given by Angstrom, we found the regression constant a and b as $a = 0.28$ and $b = 0.48$.

By getting the parameters value of δ, ω_S and $s_{\max}$ we can find the monthly average global solar radiation by calculating H_{01}. H_{01} is found maximum for the month of June which is 11.08 KWh/m^2 day and minimum for the month of December which is 6.82 KWh/m^2day. The monthly average diffused radiation is calculated by the formula given by Garg and Garg (1985). Liu and Jordan suggested to find monthly average hourly diffused radiation which lies in between the maximum and minimum value of each month calculated by the Indian Meteorological Institute.

The comparison to find total solar rays from sun on a tilted plane is done by calculating the statistical errors from the readings obtained from six empirical models which are L.J.M, K.O.M, B.A.M, H.D.M, R.E.M and H.D.K.R. The tilt was assumed to be 22 °N, and ground reflectivity is assumed to be 0.2 as for hot humid region. It is clearly visible referring from Fig. 16.1a that M.A.P.E_1 is ranging from −7.12 to −15.96% for L.J.M and R.E.M, while the other models have K.O.M −7.38%, B.A.M −5.98%, H.D.M −8.68% and H.D.K.R −8.80%.

M.B.E_1 basically provides long-term performance showing the change of real data with the measured data. From Fig. 16.1b, we can see M.B.E_1 for L.J model and K.O model is 0.28 kWh/m^2day and 0.29 kWh/m^2day , B.A model and H.D model is 0.21 kWh/m^2day and 0.36 kWh/m^2day and for R.E and H.D.K.R model is 0.81 KWh/m^2day and 0.38 KWh/m^2day.

R.M.S.E_1 provides small time working of the formulas. Ideally, it should be zero but it is good as small as possible. A large error will show high value of R.M.S.E_1. As shown in Fig. 16.1c, we can see the R.M.S.E_1 for L.U model and K.O model is 0.98 kWh/m^2day and 0.97 kWh/m^2day which is almost same. The R.M.S.E_1 for B.A and H.D is 0.97 kWh/m^2day and 1.03 kWh/m^2day and for R.E and H.D.K.R is 1.24 kWh/m^2day and 1.04 kWh/m^2day. So R.M.S.E_1 is the smallest in Badescu model (B.A.M) and the highest for Reindlet.al model (R.E.M).

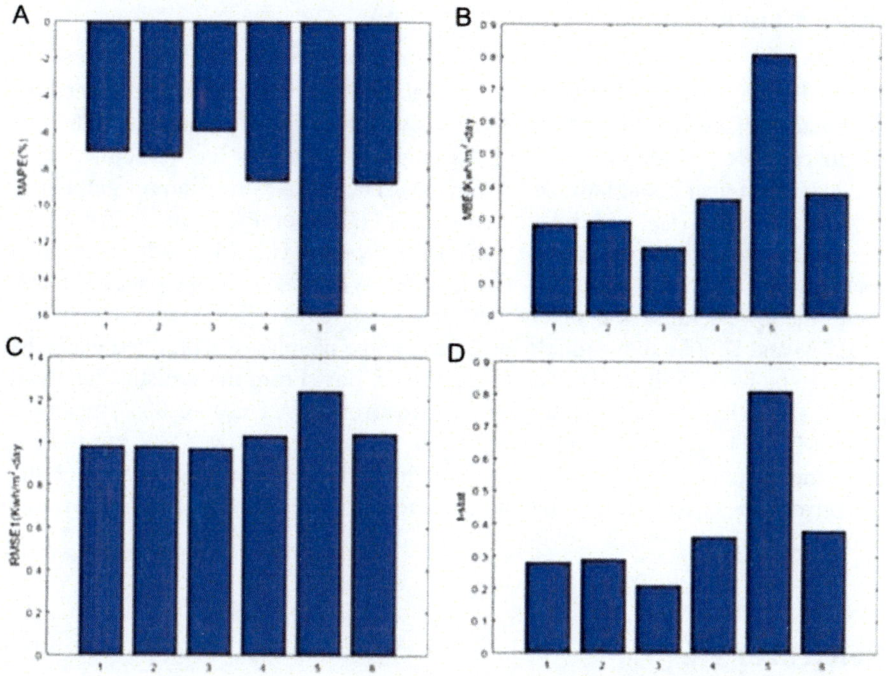

Fig. 16.1 **a** M.A.P.E$_1$, **b** M.B.E$_1$, **c** R.M.S.E$_1$ and **d** *T*-stat plotted using six empirical formulas

It is seen that M.A.P.E$_1$, M.B.E$_1$ and R.M.S.E$_1$ are not so adequate while choosing the most suitable process for measuring the solar rays on tilted plane. It is possible to have large R.M.S.E$_1$ and small M.B.E$_1$ value and vice versa. However, we can see these statistical methods provide a reasonable comparison for choosing the best model, but there is another model named *t*-stat which helps model checking and indicating whether model is statically significant. Lesser value of *t*-stat is appreciable. Referring Fig. 16.1d, we can see that the values of *t*-stat of L.J, K.O, B.A, H.D, R.E and H.D.K.R model are as follows 1.03, 1.02, 0.73, 1.23, 2.86 and 1.30. So, the smallest *t*-stat value is of B.A.M which is 0.73, and the largest is 1.30 of H.D.K.R. It is seen *t*-stat is less for isotropic than anisotropic formulas used [15]. In case of power generation from solar energy, the portion of solar rays falling on certain location has to be maximum. So, solar radiation data are taken into account for predicting the solar rays for a given location. It is simple to calculate if there is absence of elevation of the surface. But, under the presence of elevation, these models will help to calculate the sunshine over that particular area. Further, after detailed comparison, a particular model will represent or determine the maximum sunlight falling for that given area, and hence, this will lessen the time in re-calculating the radiation after few years, on the variation of climate and environment conditions and different other parameters.

16.5 Conclusion

The finding of solar rays on west side for place Vadodara, Gujarat, for inclined and horizontal surface is got with help of various variables like δ, ω_S and $s_{\max}$. The final results can be concluded by studying six empirical formulas for an inclined angle of 22° (almost near about latitude for Gujarat). The mean estimation for global rays (H_{G1}), solar rays (H_{01}) and diffused solar rays (H_{D1}) coming from sun were 6.12, 9.30 and 1.75 kWh/m^{2}day on the non-inclined plane. H.D.M concluded max. of solar rays on tilted surface that is about 6.98 kWh/m^{2}day, while Badescu model shows the lowest value of 6.38 kWh/m^{2}day.

L.J.M and K.O.M show almost same percentage of solar rays on tilted plane that seems to be 6.45 kWh/m^{2}day and 6.47 kWh/m^{2}day. From the statistical analysis, B.A.M shows less mean absolutely percentage error (M.A.P.E$_1$) −5.98%, mean bias error (M.B.E$_1$) 0.21 kWh/m^{2}day, root mean square error (R.M.S.E$_1$) and t-stat of 0.73 among six models. B.A.M shows radiation closer to measured data, and statistical error is very less so this method is preferred for estimating solar radiation in Vadodara, Gujarat.

References

1. X. Tong, C. Hu, T. Rui, W. Shen, Economical optimization for multi-microgrid based on Stackelberg Game, in *2019 14th IEEE Conference on Induatrial Electronics and Applications (ICIEA)*, pp. 1760–1764
2. M. Akmal, F. Al-Naemi, N. Iqbal, A. Al-Tarabsheh, L. Meegahapola, Impact of distribution PV generation on relay coordination and power quality, in *2019 IEEE Milan PowerTech*, pp. 1–6 (2019)
3. V.A. Shakirov, Estimation of global solar radiation and performance of photovoltaic modules using long-term meterological data, in *2018 International Multi-Conference on Industrial Engineering and Modern Technologies* (FarEastCon), pp. 1–4 (2018)
4. K. Talbi, S. Harrouni, Evaluating semi-emperical models for global solar radiation on the inclined surfaces in south of Algeria, in *2018 International Conference on Electrical Sciences and Technologies in Maghreb (CISTEM)*, pp. 1–6 (2018)
5. A.N. Eddine, I.H. Chehade, Estimation model for global solar radiation in Lebanon, in *2016 3rd International Conference on Renewable Energies for Developing Countries (REDEC)*, pp. 1–5 (2016)
6. M. A. Ansari, M. S. Anis, B. Jamil, Regression models with common coefficients for global solar radiation estimation under different climates in India, in *2018 3rd International Innovative Applications of Computational Intelligence on Power, Energy and Controls with their Impact on Humanity (CIPECH)*, pp. 86–90 (2018)
7. R. Meenal, A.I. Selvakumar, Estimation of global solar radiation using sunshine duration and temperature in Chennai, in *2016 International Conference on Emerging Trends in Engineering, Technology and Science (ICETETS)*, pp. 1–6
8. I. Baklouti, Z. Driss, M.S. Abid, Estimation of solar radiation on horizontal and inclined surfaces, in Sfax, TUNISIA, in *2012 First International Conference on Renewable Energies and Vehicular Technology*, pp. 1–10 (2012)
9. M. Kitayama da Silva, K. Barbosa de Melo, T.S. Costa, D.I. Narvez, D. de Basto, Comparative study of sky diffuse irradiance models applied in Photovoltaic Systems, in *2019 International Conference on Smart Energy Systems and Technologies (SEST)*,pp. 1–5 (2019)

10. H.S. Li, W.B. Ma, Y.W. Lian, X.L. Wang, Estimating daily global solar radiation by day of year in China. Appl. Energy **87**(10), 3011–3017 (2010)
11. B. Khalil, C. Ali, Comparative analysis of global solar radiation estimation models-a case study: Algeria, in *2019 International Conference on Advanced Electrical Engineering (ICAEE)*, pp. 1–6 (2019)
12. Solar Radiation Handbook, Solar Energy Centre, MNRE (2008), Available at http://www.indiaenvironmentportal.org.in/files/srd-sec.pdf
13. K.N. Shukla, S. Rangnekar, K. Sudhakar, Comparative study of isotropic and anisotropic models to estimate solar radiation incident on tilted surface—a case study for Bhopal, India, in *2015 Energy Reports in Elsevier*, pp. 96–103 (2015)
14. J.A. Duffie, W.A. Beckman, *Solar Engineering of Thermal Processes*, 3rd edn. (Wiley, New York, 2006).
15. A. Angstrom, Solar and terrestrial radiation. Q. J. T. Met. Soc. **50**, 121–126

Chapter 17
Implementation of PI-Controlled Converter and Monitoring of Fuel Cell on an IoT—Cloud Platform

C. Sethusubramanian, M. Vigneshpoopathy, V. Chamundeeswari, and Jayarama Pradeep

Abstract The need for renewable energy sources has increased substantially in recent years and with the technological advancements that are taking place in fuel cell technology, fuel cells may find its use in majority of the application. The advantage of using fuel cells is that it involves only electrochemical reactions to convert chemical energy to electrical energy. The work proposed here can be used for regulation of output from the fuel cell to maintain a constant output voltage since fuel cell's output is subject to variations due to the change in fuel cell parameters. A boost converter is implemented which boosts the output voltage from the fuel cell. When any variations in converter output voltage are detected, the duty cycle of the converter is changed with the help of a proportional integral (PI) controller thus bringing back the output voltage to the optimum value. The converter maintains a output voltage of 100 V which can be used in 100 V DC drives. This work projects the concept of design and implementation of fuel cell-based DC-DC converter and the implementation of Internet of things (IoT) for remote monitoring of the fuel cell parameters. The proposed work can be used majorly in automobile applications since speed control of DC motor requires a constant supply and can also be used in industrial applications involving DC drives in remote locations. Simulation of proposed circuit is executed in MATLAB/Simulink.

Keywords Boost converter · IoT · Fuel cell · Cloud platform

17.1 Introduction

In today's world, fuel cell is considered as one of the major sources of non-polluting renewable energy. The advantages of fuel cell include high efficiency conversion, not governed by any thermodynamic laws, and mainly the by product is water. Since it is pollution-free, it can be used in residential areas, built-up areas, and other places where noise-free environment is expected. Though space has its major usage of solar

C. Sethusubramanian (✉) · M. Vigneshpoopathy · V. Chamundeeswari · J. Pradeep
St. Joseph's College of Engineering, Chennai 600119, India

R. Seyezhai et al. (eds.), *Recent Trends in Renewable Energy Sources and Power Conversion*, Springer Proceedings in Energy,
https://doi.org/10.1007/978-981-16-0669-4_17

power, the other alternative is fuel cell. With these advantages, the work proposed here is to provide a constant 100 V despite of any changes in fuel flow input with the help of a boost converter and at the same time upload the fuel cell parameters such as the temperature, humidity, carbon monoxide (CO) content, fuel pressure, fuel voltage, and current to the IoT cloud platform for real-time monitoring. The fuel cells, like many other new energy sources, have the advantages, unlike the indirect conversion via heat and mechanical energy, of relatively low volume and weight, absence of moving parts, silence, and capacity to withstand overload. Also, efficiency lies between 30 and 75% and increases with decreasing load. Some of the applications of the fuel cell are provision of electric power in space vehicles, submarine propulsion and traction of a variety of surface vehicles, the provision of emergency standby electrical power sources, the maintenance of remote and seldom attended signal light and relay installations, and power storage. The use of fuel cell in association with technical electrolytic process to economize energy and the utilization of low-grade fuels of biological origin are additional applications [1, 2].

Section 17.2 deals with explanation of the proposed system, Sect. 17.3 deals with the implementation of the proposed system, Sect. 17.4 deals with the Hardware implementation of the proposed converter, and Sect. 17.5 concludes the work.

17.2 Proposed System

In this work, a system is proposed for the fuel cell which maintains a constant DC voltage at the output with the help of a DC-DC boost converter with the implementation of proportional integral (PI) controller that varies the duty cycle of the converter to maintain the required constant output voltage [3, 4]. Arduino Uno microcontroller is used as PI controller [5, 6] and NodeMCU microcontroller to send the data collected from various sensors to the IoT cloud platform.

Figure 17.1 shows the proposed system, and the respective components of the proposed system are explained in the following section.

17.2.1 Fuel-Cell Stack

Fuel cell stack that is shown in Fig. 17.1 is a 6 kW 45 V polymer electrolyte membrane (PEM).

Fuel cell stack has inputs as hydrogen and air. The fuel cell provides a 45 V DC output which may vary with the fuel cell parameters [7]. A PEM fuel cell stack is considered in this, as it offers higher efficiency when compared to other fuel cells [8]. The characteristics of the fuel cell stack are shown in Fig. 17.2.

Figure 17.2 portrays the rating of the fuel cell in terms of its voltage and power as 45 V and 6 kW.

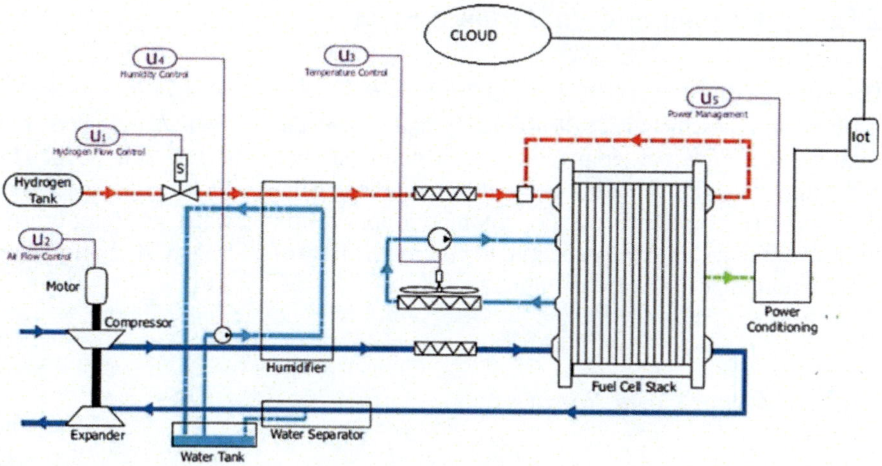

Fig. 17.1 Block diagram of the proposed system

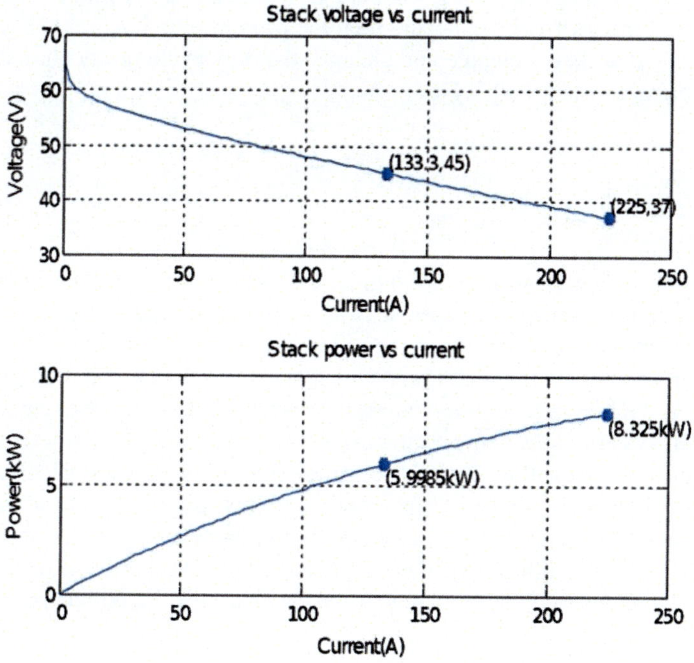

Fig. 17.2 Characteristics of fuel cell stack

17.2.2 Air Flow and Fuel Flow System

Operation of the fuel cell stack at high pressure improves the reaction rate which further improves the fuel cell efficiency and power density [9]. An air cooler is required to reduce the temperature of air before it enters the fuel cell stack. The humidifier unit adds vapor into the air flow as indicated in Fig. 17.1 in order to prevent dehydration of the membrane. For automotive applications, a water separator is used to recover the water. A valve is used to control the flow of hydrogen. A humidifier is used to humidify the hydrogen flow.

17.2.3 Power Conditioning

In order to provide a constant DC output supply, a boost regulator chopper is used to boost the input to 100 V DC since DC motors of high power applications use 100 V input. The output from the chopper is sensed using a voltage sensor and if any deviation from set voltage occurs then the proportional integral (PI) controller automatically detects the change and changes the duty cycle of the chopper so that output remains constant 100 V DC.

17.2.4 Sensors

The sensors used in the proposed system are carbon monoxide (CO) sensor, temperature and humidity sensor, pressure sensor, current and voltage sensor. CO sensor is used to detect the CO content in the membrane since excess of CO content can cause CO poisoning of the membrane which results in the degradation of fuel cell efficiency in the long run. Temperature sensor is used to measure the temperature of the fuel cell system. If the temperature exceeds the nominal operating temperature, then cooling systems will be initiated. Similarly humidity sensor is used to measure the humidity content in the input of the fuel cell. Pressure sensor is used to measure the input pressure of the fuel and air since fuel pressure is one of the main factors that affect the output voltage of the fuel cell. The current sensor measures the output current of the converter, and the voltage sensor measures the output voltage of the converter.

17.2.5 IoT System

The IoT system uses a NodeMCU module that collects the data from various sensors and regularly updates these values to the cloud platform which can be then used for real-time monitoring of the fuel cell [10]. NodeMCU module uses ESP8266 microcontroller which is programmed to receive the sensor outputs using the analog pin of the microcontroller. Since only one analog pin is present in the NodeMCU, a multiplexer (CD74HC4067) is used to obtain analog signals from various sensors and sends it to the NodeMCU. The NodeMCU is programmed to connect to a Wi-Fi with an active Internet service. The sensor data is then sent to the cloud platform's channel using an API key which is unique for that channel. The NodeMCU module is programmed with the help of Arduino IDE.

17.2.6 Cloud Platform

The cloud platform used in this work is ThingSpeak; this is an open source cloud platform with the provision to include MATLAB analysis.

17.3 Implementation of the Proposed System

The boost converter fed by a fuel cell is considered in this work and is simulated using MATLAB and the performance of the converter is studied by introducing change in the fuel flow rate (55–85 lpm) with the help of a flow rate regulator.

Figure 17.3 shows the Simulink model of the proposed system and Fig. 17.4 shows the DC-DC boost converter subsystem. The simulation parameters of the fuel cell are shown in Table 17.1.

Table 17.2 shows the simulation parameters of the DC-DC boost converter.

The performance parameters of the fuel cell is shown in Fig. 17.5.

Figure 17.5a shows the variation in flow rate of hydrogen. The hydrogen flow is increased from 50 to 85 lpm in 3.5 s to study the performance parameters of the converter as well as the fuel cell. Figure 17.5b shows the variation in the % utilization of O_2 and H_2. Figure 17.5c shows the variation of air and fuel consumption of the stack. Figure 17.5d shows the variation of stack efficiency of the fuel cell due to the change in hydrogen flow rate.

The output voltage and current waveforms of the fuel cell before implementation of the converter are shown in Fig. 17.6.

Figure 17.6a, b shows the variation in the fuel cell output voltage and current when hydrogen flow is increased from 50 to 85 lpm. At 50 lpm, it is observed that the fuel cell output is 45 V but as the fuel flow increases to 85 lpm the fuel cell output is 54.5 V. The converter has to provide a constant 100 V so at 50 lpm duty ratio is calculated from the output equation of the boost converter, as in (17.1):

$$V_s = V_o(1 - D) \tag{17.1}$$

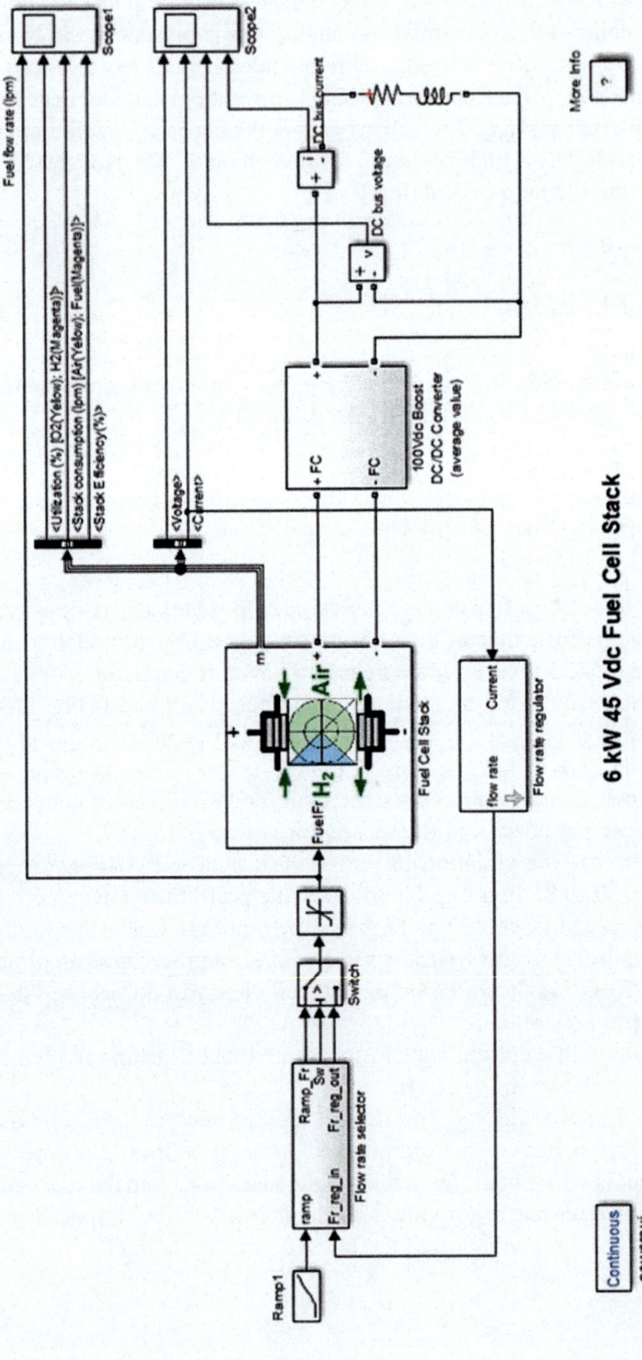

Fig. 17.3 Simulink model of the system proposed

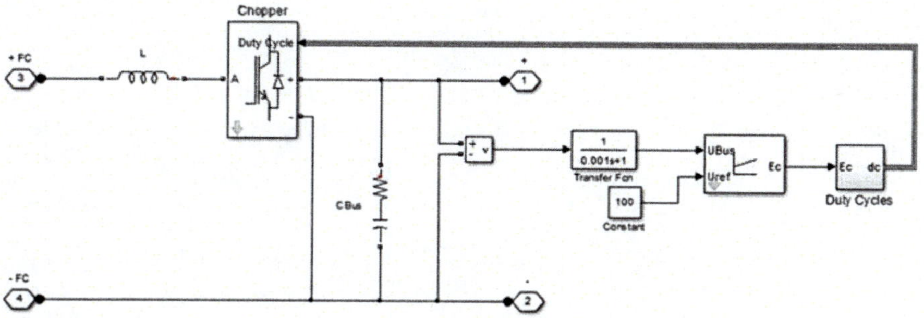

Fig. 17.4 Simulink model of converter

Table 17.1 Simulation parameters of the fuel cell stack

Parameter	Value
Power rating	6 kW
Nominal operating voltage	45 V
Nominal operating current	133.3 A
Maximum operating point	225 A
Voltage at maximum operating piont	37 V
Number of cells	65
Nominal stack efficiency	55%
Operating temperature	65 °C
Nominal air flow	300 lpm
Nominal supply pressure (fuel)	1.5 bar
Nominal supply pressure (air)	1 bar

Table 17.2 Simulation parameters of the boost converter

Parameter	Value
Input voltage	45 V
Switching frequency	62 kHz
Inductance (*L*)	500 μH
Capacitance (*C*)	7500 μF
Load resistance (R)	1.6667Ω

Here Vs is the fuel cell output and Vo is the converter output. Thus, it is observed the duty ratio changes from 0.55 to 0.455. This change is done using the PI controller. Equations for the proportional mode and integral mode are combined, to have an analytic expression for this mode, which is given as in (17.2):

$$P = K_p e_p + K_i \int e \int e_p \, dt + pt(0) \tag{17.2}$$

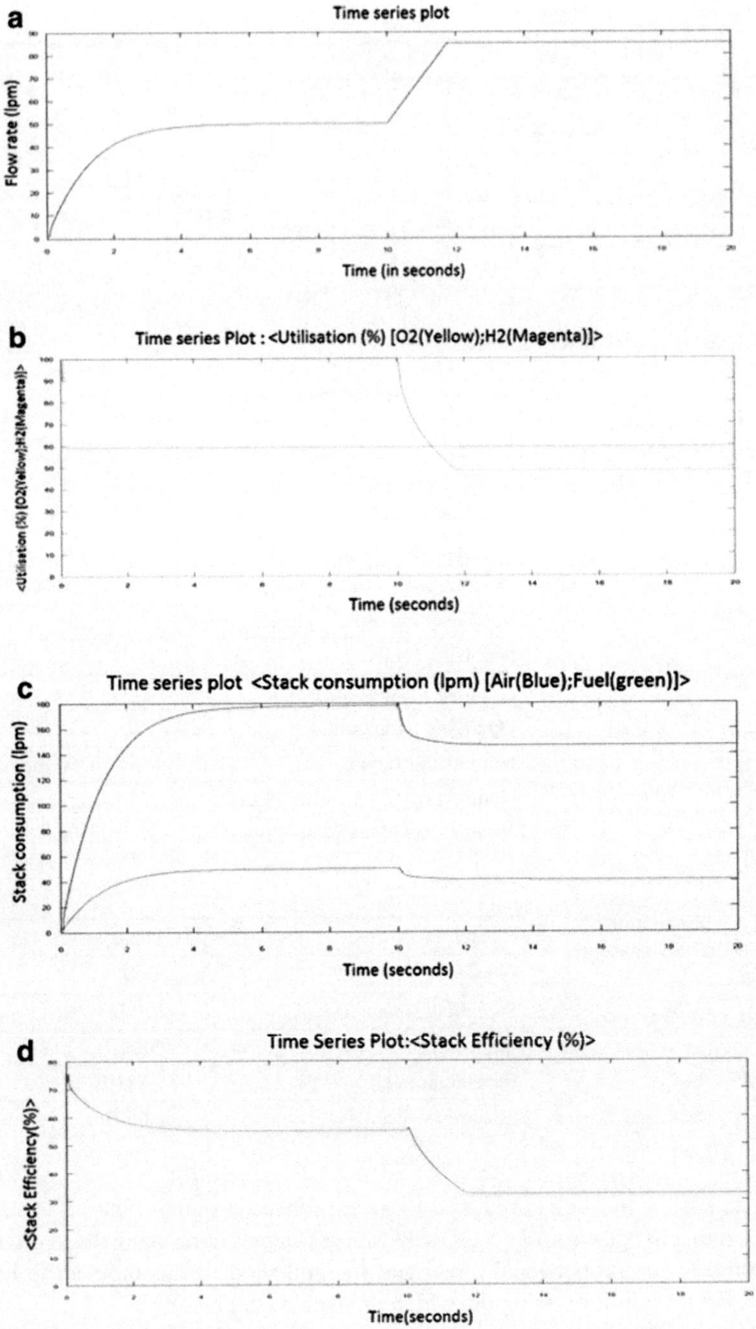

Fig. 17.5 **a** Hydrogen flow rate, **b** utilization of oxygen and hydrogen, **c** air and fuel consumption, **d** stack efficiency

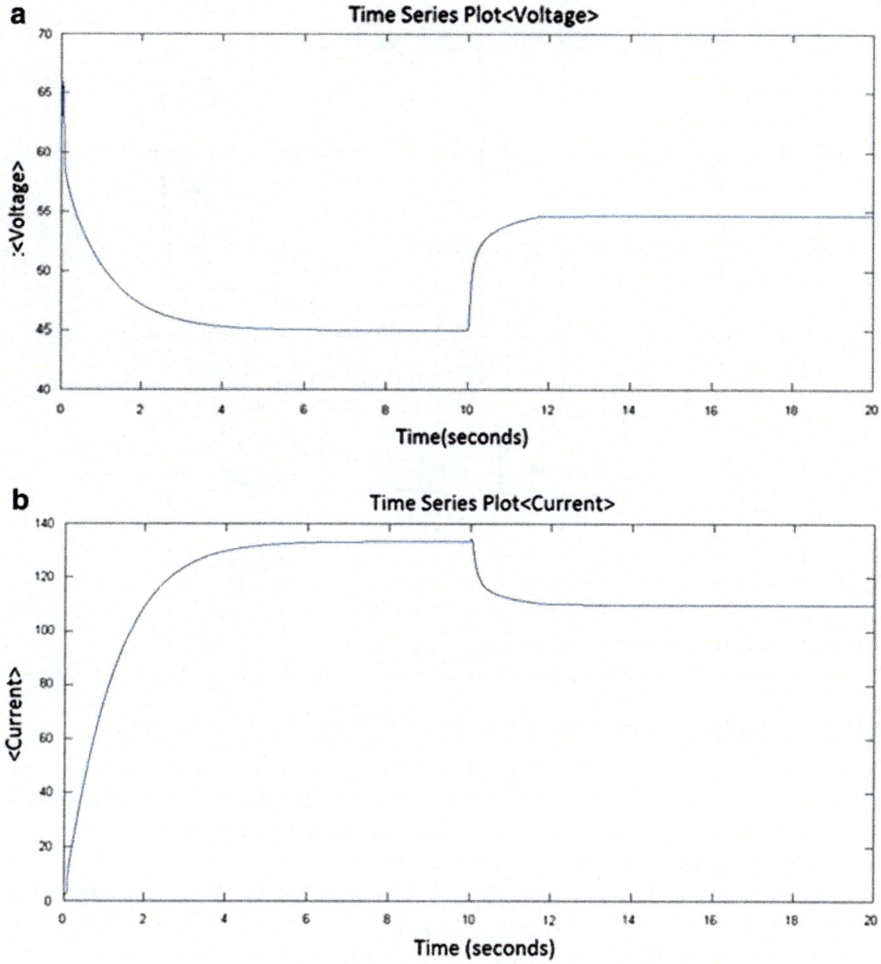

Fig. 17.6 **a** Output voltage waveform of fuel cell, **b** output current waveforms of fuel cell

The PI implementation on converter is shown in Fig. 17.7. The constants *K*p and *K*i are calculated practically such that the output reaches steady state within lesser duration.

After the implementation of the boost converter, the output becomes a constant 100 V as shown in Fig. 17.7.

Figure 17.8a, b shows the output voltage and current waveform with the implementation of proportional integral (PI) controller boost converter. It can be seen from Fig. 7a that the output is maintained at 100 V despite of the changes in the hydrogen input flow rate.

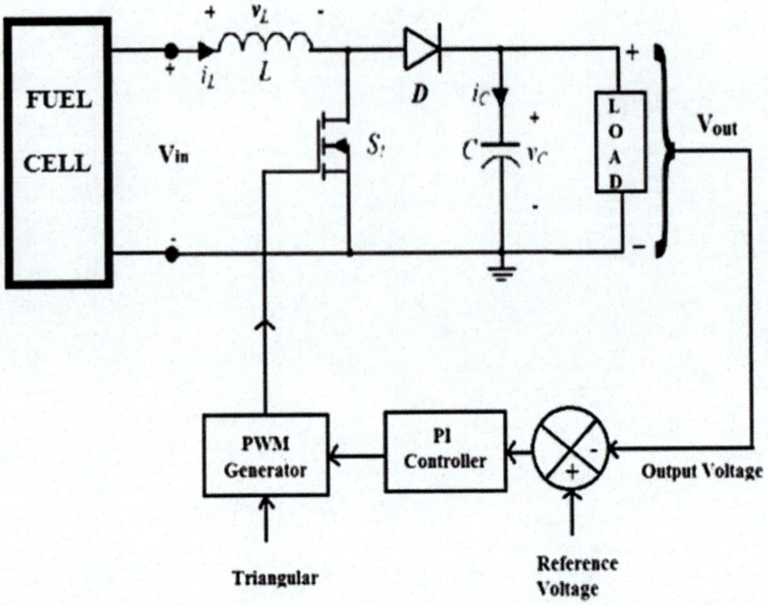

Fig. 17.7 PI implementation

17.4 Hardware Implementation of Proposed Converter

The prototype model of the proposed converter was designed and the input to the converter was varied to check if can provide a constant output. Figure 17.9 shows the hardware implementation of the converter.

The IoT interface used for this paper is ThingSpeak which is used to monitor the fuel cell parameters on remote systems. The monitoring of sample data on the ThingSpeak cloud platform is shown in Fig. 17.10.

17.5 Conclusion

In this paper, a study on fuel cell system is made and an improved system is proposed for maintaining a constant output voltage and also monitoring the fuel cell parameters. The simulation of the proposed system is modeled using MATLAB and the fuel cell output variations were studied for change in input fuel flow. Based on this study, a PI-controlled boost converter was designed to maintain the output as constant. The output waveforms of the converter were studied, and it was found that the converter maintains a constant DC output. ThingSpeak cloud platform was used to monitor the data from the sensors. Hardware model of the proposed system was implemented successfully. This system can be implemented to minimize the variations in the fuel

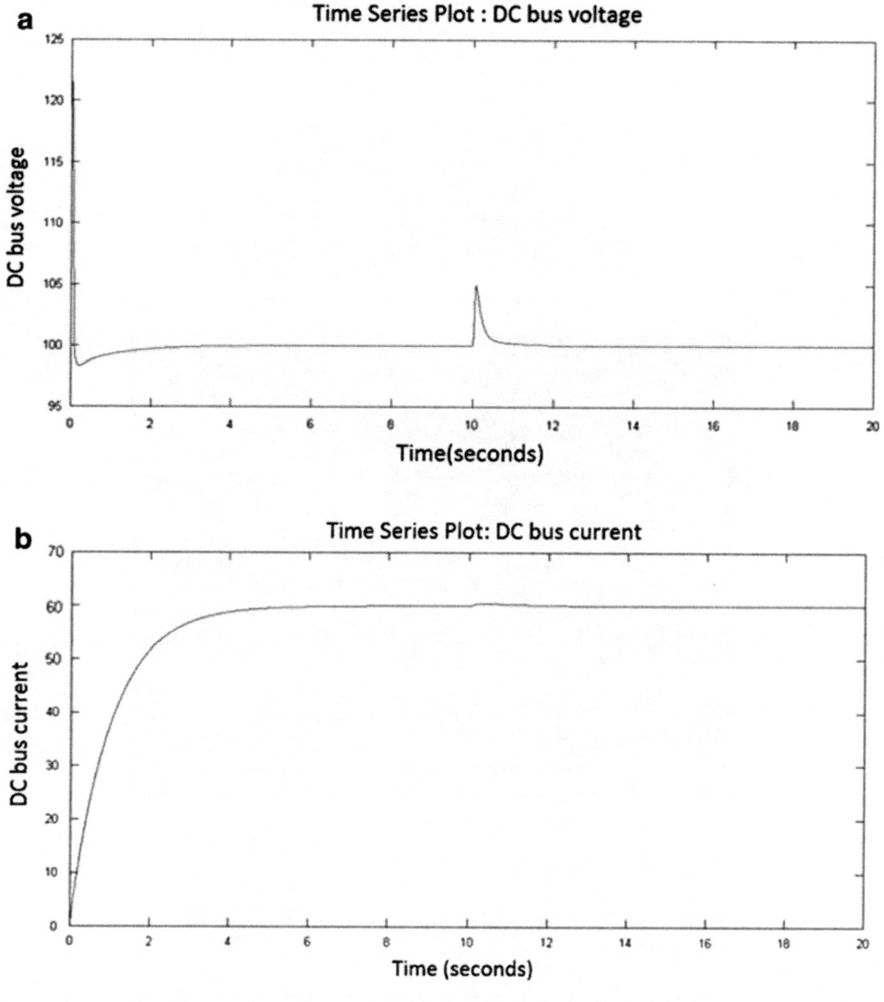

Fig. 17.8 **a** Output voltage waveform of the converter, **b** output current waveforms of the converter

cell output due to change in fuel flow and to remotely monitor the parameters of the fuel cell system.

17.6 Future Scope

This paper demonstrates the use of PI-controlled boost converter and monitoring systems for fuel cell applications. With the technological advancements in fuel cells, such systems can be implemented in fuel cell vehicles and can be integrated easily

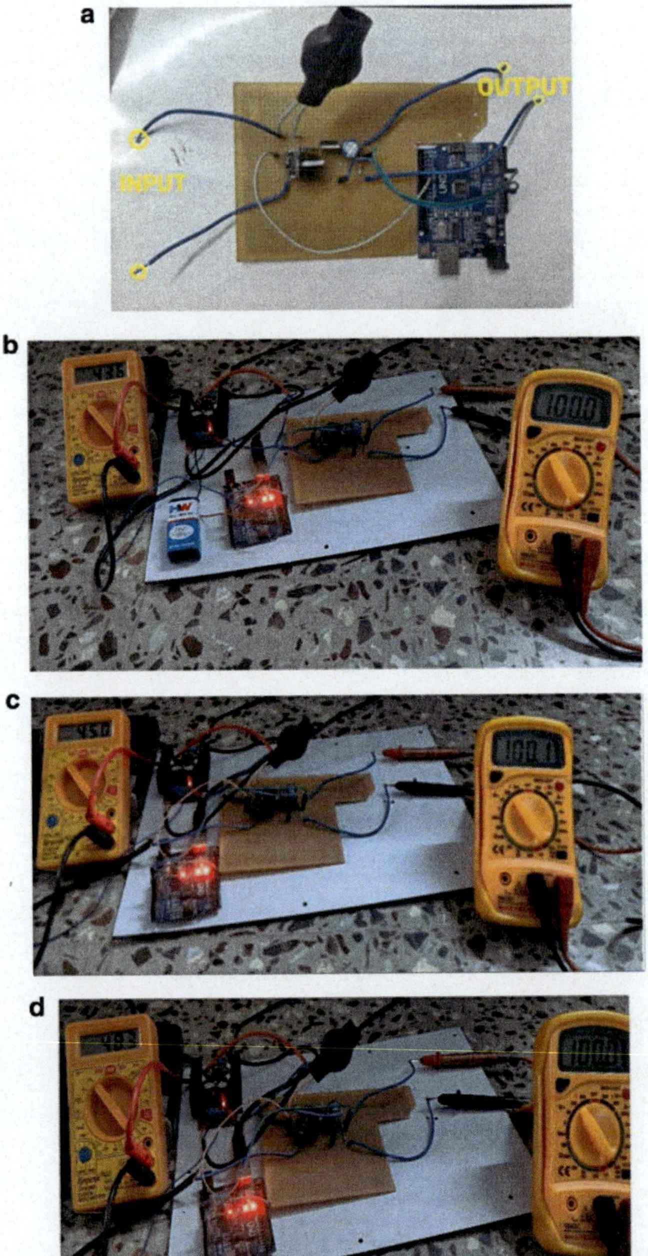

Fig. 17.9 **a** Hardware implementation of converter, **b** input voltage = 43.6 V, output voltage = 100.0 V, **c** input voltage = 45 V, output voltage = 100.1 V, **d** input voltage = 48.3 V, output voltage = 100.0 V

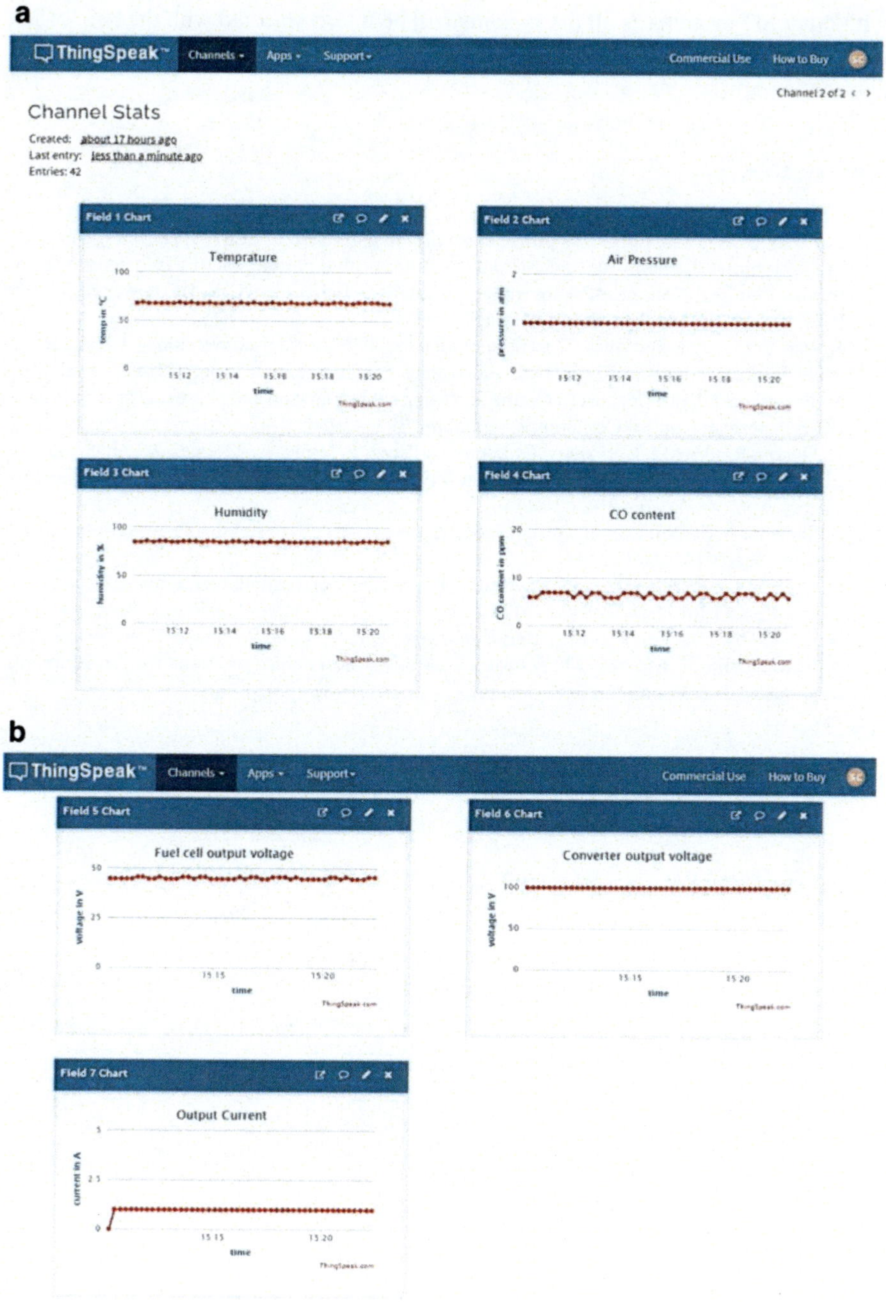

Fig. 17.10 **a** Fuel cell parameters, **b** converter parameters

with other IoT systems as all the systems will be interconnected with the help of IoT in near future.

References

1. J.-S. Lai, M.W. Ellis, Fuel-cells power systems and applications. IEEE Spectrum (2017)
2. M. LaMonica, GE claims fuel cell breakthrough starts pilot production. IEEE Spectrum (2014)
3. H. Sira-Ramirez, Nonlinear P-I controller design for switchmode DC-to-DC power converters. IEEE Trans. Circuits Syst. **38**(4), 410–417 (1991)
4. K.Wan, J. Liao, M. Ferdowsi, Control methods in DC/DC power conversion—A comparative study. In: Proceedings of IEEE power electronics conference, vol. 1, pp. 921–926, Jun 2007
5. M. Nagurka, O. Yaniv, Robust PI controller design satisfying gain and phase margin constraints. IEEE American Control Conference, vol. 5, pp. 3931–3936, 2003.
6. J. Alvarez-Ramirez, I. Cervantes, G. Espinosa-Perez, P. Maya, A. Morales,A stable design of PI control for DC-DC converters with an RHS zero. IEEE Trans. Circuits Syst. I Fundam. Theory Appl. **48**(1):103–106 (2001)
7. A. Samson, A. Senthil Kumar, Design and analysis of fuel cell for standalone re-newable energy system, Dec 2014
8. H.W. Zhang, P.K. Shen, Recent development of polymer electrolyte membranes for fuel cells. Chem. Rev. **112**(5), 2780–2832 (2012)
9. A.J. Appleby, F.R. Foulkes, *Fuel Cell Handbook* (Van Nostrand Reinhold, New York, 1989).
10. M.S. Radicella, R. Struzak, M. Zennaro, Educating on wireless solutions for environmental monitoring. J. Telecommun. Inf. Technol. (JTIT), **4**, 78–82 (2012)

Author Index

R. Seyezhai et al. (eds.), *Recent Trends in Renewable Energy Sources and Power Conversion*, Springer Proceedings in Energy,
https://doi.org/10.1007/978-981-16-0669-4